SpringerBriefs in Immunology

More information about this series at http://www.springer.com/series/10916

Sheikh Rayees • Inshah Din

Asthma: Pathophysiology, Herbal and Modern Therapeutic Interventions

Springer

Sheikh Rayees
Department of Pharmacology
University of Illinois at Chicago
Chicago, IL, USA

Inshah Din
Department of Biochemistry
Government Medical College
Srinagar, Jammu and Kashmir, India

ISSN 2194-2773   ISSN 2194-2781   (electronic)
SpringerBriefs in Immunology
ISBN 978-3-030-70269-4   ISBN 978-3-030-70270-0   (eBook)
https://doi.org/10.1007/978-3-030-70270-0

This Springer imprint is published by the registered company Springer Nature Switzerland AG
The registered company address is: Gewerbestrasse 11, 6330 Cham, Switzerland

Introduction

Asthma is a global health problem causing enormous mortality and morbidity, both in developed and developing countries. Despite remarkable advances in diagnosis and treatment, asthma is still a serious public health problem, particularly due to the off targets and side effects of the commonly available drugs. Asthma has shown a drastic increase in global prevalence, morbidity, mortality, and economic burden over the last 40 years, especially in children (Beuther 2010; Masoli et al. 2004).

Allergic asthma is a common chronic lower airway disease. It is mediated by a dominant T helper 2 (Th2) immune response and its distinctive features include airway hyperresponsiveness (AHR), increased immunoglobin (IgE) levels, airway inflammation, and mucus hypersecretion. An allergic immune response is generated upon stimulation of T cell receptors by antigen ligation which results in differentiation of peripheral CD4$^+$ T cells into effecter T cells. These cells are classified as Th1, Th2, Th17, and Tregs (regulatory T cell) based on their cytokine signature. Th2 lymphocytes producing a type 2 cytokine profile are known to mediate pathophysiology of asthma. These cells are necessary to allergic asthma development in both animals and humans. The amount of type 2 cytokines produced by them is highly elevated in the airway tissues of human asthma subjects and animals in which asthma is induced, using suitable allergens (Gavett et al. 1994; Robinson et al. 1992). Individually or synergistically, Th2 cytokines mediate vital functions in asthma, e.g., Interleukin-5 (IL-5) is known to play an imperative role in eosinophil maturation, differentiation, recruitment, and survival. IL-4 along with IL-13 is essential in eosinophil accumulation and is a primary factor in IgE production by B cells. IL-13 is however known to play the most important and central role in allergic asthma. This Th2 cytokine alone has been found to be sufficient and necessary for the induction of allergic asthma in mice (Mosmann et al. 1986; Cohn and Ray 2000). It contributes in airway fibrosis, airway hyperresponsiveness, mucus production, IgE synthesis, and airway inflammation (Wills-Karp and Karp 2004). Interleukin-4 is responsible for Th2 cell differentiation which is executed after IL-4 binds to its receptor causing phosphorylation and dimerization of a signaling protein known as signal transducer and activator of transcription 6 (STAT6), a Th2 cell-specific transcription factor, through the JAK/STAT cascade. This modulated form of STAT6 translocate from cytoplasm to the nucleus to activate the transcription of cytokine responsive genes, especially GATA3 (Zhu et al. 2001). GATA3 activates the transcription of IL-5 and IL-13 genes by directly binding to their promoters (Zhou and Ouyang 2003). Moreover, this transcription factor is also involved in remodeling of chromatin structure and opening of IL-4 locus (Ouyang et al. 2000).

The treatment of asthma by herbal medicine has been practiced since ages throughout the world. There are many herbal therapies which are in practice globally (Saganuwan 2010). The herbal therapies include all sorts of medicine be it anti-histaminic, smooth muscle relaxant, or anti-inflammatory. Doubtlessly, the herbal medicines have been able to treat asthma pathologies like airway hyperresponsiveness, airway inflammation, or smooth muscle relaxation. Additionally, some plant-based drugs were developed over time against several respiratory ailments including asthma (Lee 2000; Graham and Blaiss 2000; Bielory and Lupoli 1999).

Currently available drugs and therapies for asthma involve inhaled corticosteroids, β2-adrenoceptor agonists, leukotriene modifiers, phosphodiesterase inhibitors, cytokine-based immunotherapies, and transcription factor modulators. However, none of these drugs or therapies is individually sufficient to treat any form of asthma whether mild, moderate, or severe because of asthma heterogeneity, steroid resistance, and poor biological half-life of β2-adrenoceptor agonists. So, the need for newer and novel therapies is warranted (Beuther 2010; Masoli et al. 2004; Wills-Karp and Karp 2004; Zhu et al. 2001).

In this book, we will try to review all the literature related to pathophysiology of asthma and its modern and herbal medicine-based therapies.

Contents

Chapter 1
Asthma

Asthma is a chronic airway disease affecting over 300 million people worldwide, including 10% (30 million) in India, with an expected increase of a further 100 million by 2025 (Bel 2013; Kant 2013). Past decade has observed a notable increase in asthma prevalence on both national and global levels with highest rates observed in western countries (about 30%). So, attributing this rise exclusively to air pollutants is not genuine, as the air quality in these countries is better than Delhi, an Indian metropolitan city, where the rate is about 10% (Agrawal and Gosh 2011). Over the past 40 years, a drastic increase in global prevalence, morbidity, mortality, and economic burden have been observed due to asthma especially in children (Beuther 2010; Masoli et al. 2004). The rising numbers of hospital admissions for asthma, especially young children, reflect an increase in severe asthma, poverty, and lack of proper disease management. Worldwide, approximately 180,000 deaths annually are caused due to asthma. The financial burden on a single asthma patient per year in different western countries ranges from US$300 to 1300 (Braman 2006).

Asthma is a chronic airway inflammatory disorder, characterized by airway hyperresponsiveness, bronchoconstriction, reversible airflow obstruction, and airway inflammation. Airway hyperresponsiveness is a pathological situation of airways where they become hyperresponsive to a stimulus whether inhaled constrictor or an agonist, resulting in wheezing, chest tightness, and coughing (Sporik et al. 1995). The characteristic features of asthma include airway inflammation, airway remodeling, mucus hypersecretion, and chronic elevation of serum IgE, which is intimately associated with allergy (Ray and Cohn 1999).

S. Rayees, I. Din, *Asthma: Pathophysiology, Herbal and Modern Therapeutic Interventions*, SpringerBriefs in Immunology, https://doi.org/10.1007/978-3-030-70270-0_1

Chapter 2
Types of Asthma

Asthma may be categorized as atopic (extrinsic) or non-atopic (intrinsic), based on whether symptoms are precipitated either by allergens such as house dust mite and pollens (atopic) or not (non-atopic). However, on the basis of severity, asthma is classified as:

2.1 Mild to Moderate Asthma

The clinical characterization of this form of asthma mainly involves a 20–40% reduction in FEV1 (forced expiratory volume) with a 20–30% variability. Pathologically, it involves acute to chronic inflammation in airways consisting of infiltrated eosinophils, activated Th2 lymphocytes, elevated IgE levels, and goblet cell hyperplasia causing AHR and remodeling of the airway epithelium. Upregulation of Th2 cytokines especially IL-4, IL-5, and IL-13 contribute to various pathological features of the disease (Shikotra et al. 2012).

2.2 Severe Asthma

Severe asthma is clinically differentiated from other forms of asthma by reduction of more than 60% FEV1 with >30% variability. The pathological features of severe asthma include elevated levels of lymphokines such as IFN-γ, IL-27, TNF-α, and IL-17 and a diverse granulocyte airway infiltrate. The patients with severe asthma have a mixed Th2/Th1 phenotype and are resistant to glucocorticoid treatment (Von Bulow et al. 2014; Knarborg et al. 2014).

© The Author(s), under exclusive licence to Springer Nature Switzerland AG 2021
S. Rayees, I. Din, *Asthma: Pathophysiology, Herbal and Modern Therapeutic Interventions*, SpringerBriefs in Immunology,
https://doi.org/10.1007/978-3-030-70270-0_2

2.3   Pathophysiology Asthma

Th2 lymphocytes (CD4+ Th2 cells) secreting type 2 cytokines are known to mediate pathophysiology of asthma. However, other major cellular components are also involved which include eosinophils, mast cells, dendritic cells, and macrophages (Barrett and Austen 2009). CD4+ T cells are crucial in controlling airway inflammation in asthma. They infiltrate the asthmatic airways, get activated by antigens, and express MHC class II and CD25 (IL-2R) surface activation markers (Corrigan et al. 1988). These antigen-activated cells differentiate into effecter cells (T cell subsets), which are then recognized on the basis of cytokines they release. These include Th1, Th2, Th17, and T regulatory cells (Cher and Mosmann 1987; Kim et al. 1985). Th17 cells promote neutrophil recruitment to the airways during asthma for the clearance of bacteria and fungi (Cosmi et al. 2011), whereas regulatory T cells modulate the immune system, repeal autoimmune diseases, and maintain tolerance to self-antigens (Hori et al. 2003). Th1 cells produce macrophage-activating factor, IFN-γ and lymphotoxin, which mainly incite a strong cell-mediated immune response, mainly against intracellular pathogens, whereas Th2 cells produce a different panel of cytokines, including IL-4, IL-5, IL-9, IL-10, and IL-13. These cytokines are known to mediate the pathophysiology of asthma. Th2 cells are necessary to allergic asthma development in both rodents and humans. Among the cytokines released by them, IL-4, IL-5, and IL-13 are of prime importance in mediating most of the asthma features (Mosmann et al. 1986).

2.4   Interleukin-5

This Th2 cytokine is essential for differentiation, maturation, and survival of eosinophils. It also stimulates the growth of B cells and augments immunoglobulin secretion through its binding to IL-5 receptor (Ichinose and Barnes 2004; Walsh 2011).

2.5   Interleukin-13

IL-13 is known as a central mediator of allergic asthma. Alone, IL-13 has been found to be sufficient and necessary for allergic asthma induction in rodents and airway inflammation development in humans (Anderson 2008). It plays critical role in AHR development, eosinophil production, airway fibrosis, and mucus production. IL-13 is also necessary for IgE production, along with IL-4 (Fallon et al. 2001; Mckenzie et al. 1998).

2.6 Interleukin-4

This cytokine is responsible for differentiation of naive helper T cells to Th2 cells (Sokol et al. 2008). Most of its biological functions are similar to IL-13 as both share same receptor complex of IL-13Rα1/IL-4Rα (type II) (Ramalingam et al. 2008). IL-4 and IL-13 are responsible for inducing fibrotic changes in asthmatic lungs and causing eosinophil accumulation. Furthermore, IL-4 is involved in generation of first signal to IgE switching (Mckenzie et al. 1998; Webb et al. 2000).

Chapter 3
Airway Inflammation and Airway Hyperresponsiveness

Airway inflammation is a complex inflammatory process of airways involving the contribution of various cells, recruited and activated in asthmatic conditions, and the mediators released by them. However, major components contributing to this inflammation are eosinophils, mast cells, and lymphocytes. In inflamed airways, the lumen becomes narrow and edematous due to mucus plug formation, made up of proteins from airway vessels and epithelial cells of airways. Airway inflammation also involves epithelial wall shedding that has been observed in the airway lumen of asthmatic subjects. These changes are also observed in patients suffering from mild asthma (Busse et al. 2001; Bousquet et al. 2000).

Airway hyperresponsiveness is regarded as a cardinal feature of asthma and is associated with various symptoms of asthma like wheezing, chest tightness, and bronchoconstriction. A relationship between degree of hyperresponsiveness and inflammation of airways has been reported in various studies. Inflammation of the airways causes an increase in airway hyperresponsiveness and various other symptoms like cough, chest tightness, and wheezing by stimulation of airway sensory nerve endings. Still, mechanisms involved in airway inflammation are yet to be explored properly (Barnes 1996; Bousquet et al. 2000; Toward and Broadley 2002).

S. Rayees, I. Din, *Asthma: Pathophysiology, Herbal and Modern Therapeutic Interventions*, SpringerBriefs in Immunology, https://doi.org/10.1007/978-3-030-70270-0_3

Chapter 4
Effects of Inflammation

Airway inflammation is associated with various structural alterations in airways termed as airway remodeling, recruitment, and activation of inflammatory cells and impaired lung functioning. These processes result in various pathophysiological changes associated with asthma (Fig. 4.1). However, in many asthma patients, this does not alter the proper lung functioning, which depicts the role of genetic factors in progression of these features (Lange et al. 1998; Ulrik and Lange 1994; Ryu et al. 2014). Airway inflammation results in epithelial wall shedding, fibrosis, mucus hypersecretion, and abnormalities in airway smooth muscle functioning.

4.1 Effects on Airway Epithelia

Airway inflammation results in epithelial wall shedding which is one of the distinctive features of asthma. The epithelia is shed due to various mediators like eosinophil peroxidase, eosinophil cationic protein, leukotrienes like leukotriene C4, D4, and E4 and oxygen-derived free radicals, released by inflammatory cells. This damage in epithelial wall contributes to the development of AHR in several ways including loss of activity of enzymes that degrade inflammatory mediators (e.g., neutral endopeptidase) and sensory nerve exposure leading to reflex neural effects (Knight et al. 2001; Holgate et al. 2000).

4.2 Fibrosis

The basement membrane of asthmatic airway epithelium deposits Type III and V collagen, causing subepithelial fibrosis (Redington 2000; Silva et al. 2014; Chetta et al. 1997). The role of fibrosis is scantily understood in asthma. Fibrosis has been

S. Rayees, I. Din, *Asthma: Pathophysiology, Herbal and Modern Therapeutic Interventions*, SpringerBriefs in Immunology,
https://doi.org/10.1007/978-3-030-70270-0_4

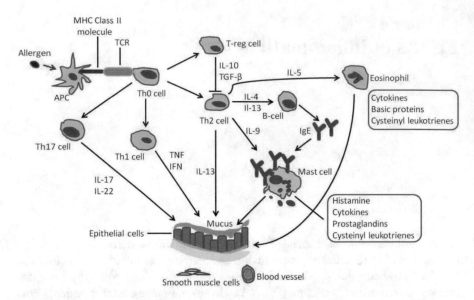

Fig. 4.1 Various interacting inflammatory cells and mediators, which contributing in asthma pathophysiology. The binding of allergen to antigen-presenting cells leads to their activation which then interacts with the Th cells. Th0 cell is the mother of all Th cells, be it T regulatory cells, Th1 cells, Th17 cells, or Th2 cells. One of the most important cells demonstrated above are the Th2 cells. They interact with other nearby cells through their secreted cytokines such as IL-9, IL-4, IL-5, and IL-13. These cytokines are central to the development of allergic asthma. They exert their effect by interacting/activating other cell types like mast cells, B cells, eosinophils, etc. The activation of these cells and resultant secreted mediators further exaggerate the asthma phenotypes

observed in mild to severe forms of this disease in airway smooth muscle and in lower strata of epithelial cells. It has been reported that fibrosis causes irreversible loss of lung function which becomes worse in persistent uncontrolled asthma. So, a role of fibrosis in asthma is likely (Chetta et al. 1997; Kips and Pauwels 1999; Fish and Peters 1999).

4.3   Mucus Hypersecretion

Mucus hypersecretion is one of the primary features of asthma. An enhanced mucus secretion causes the formation of a mucus plug in asthmatic airways. This is caused by amplification in the number of goblet cells (hyperplasia) and an increase in size of submucosal glands (hypertrophy) (Ordonez et al. 2001; Toward and Broadley 2002). The increased secretion of mucus occurs as a result of Th2 cytokines and epithelial growth factor (EGF) by modulation of mucin gene muc5AC in both animal models and humans. Mucus hypersecretion contributes in AHR, narrowing of the air passage, wheezing, and chest tightness (Zhou et al. 2014; Ordonez et al. 2001; Zhu et al. 1999; Nakanishi et al. 2001; Farahani et al. 2014).

4.4 Airway Smooth Muscle

Airway smooth muscle, present in the trachea and bronchial tree, is an important tissue associated with bronchomotor tone and therefore is important in asthma pathophysiology. However, airway smooth muscle of asthmatic patients usually does not show a good response to spasmogens under in vitro conditions. Further, this tissue shows a non-significant response to β-adrenergic agonists when removed surgically from asthmatic patients without reducing the number of β-receptors in postmortem (Barnes 1998a, b; Petrovic-Stanojevic et al. 2014). Furthermore, in vitro and in vivo stimulation of airway smooth muscles to inflammatory cytokines (e.g., IL-1β), downregulates their response to β-adrenergic agonists. There are various other reports that clearly advocate a concrete link between airway inflammation and smooth muscle functioning in asthma. The regulation of bronchial tone by airway inflammatory processes may be due to modulation of ion channels of airway smooth muscles which regulate their resting membrane potential and hence change their level of excitability (Grainge et al. 2011; Mak et al. 2002; Grandordy et al. 1994; Laporte et al. 1998).

Chapter 5
Inflammatory Cells Involved in Asthma

There are a variety of cells that contribute in development and persistence of airway inflammation and hence asthma. However, the main cells involved are eosinophils, mast cells, dendritic, macrophages, and CD4+T Cells:

5.1   Eosinophils

Eosinophils are major inflammatory cells involved in allergic diseases. Eosinophils are not normally present in healthy lungs, but their accumulation in lungs marks an inflammatory condition of airways, in mild to severe forms of asthma (Walsh et al. 2005). Eosinophils are white blood cells and are involved in various inflammatory processes during allergic asthma. Essentially IL-5 is responsible for their maturation, survival, and recruitment to the airways. Eosinophils contribute by releasing several cationic proteins especially eosinophil peroxidase and reactive oxygen species like hypobromite, superoxide, and peroxide. They also release lipid mediators such as leukotrienes like LTC4 and LTE_4 and prostaglandins, e.g., PGE_2. Furthermore, eosinophils secrete cytokines such as IL-1, IL-2, IL-4, IL-5, IL-6, IL-8, IL-13, and TNF alpha and growth factors such as TGF beta, VEGF, and PDGF. The accumulation of eosinophils in the lungs causes inflammation in upper and lower airways. The potent pro-inflammatory products, especially eosinophil peroxidase and leukotrienes, released by them are known to cause remodeling of epithelial wall in asthmatic airways (Blanchard et al. 2005).

S. Rayees, I. Din, *Asthma: Pathophysiology, Herbal and Modern Therapeutic Interventions*, SpringerBriefs in Immunology, https://doi.org/10.1007/978-3-030-70270-0_5

5.2   Mast Cells

These are resident granulocytes present in different tissues of the body containing granules, rich in histamine and heparin. Mast cells are found in significantly increased numbers in airway smooth muscles of asthma patients. These cells cause bronchoconstriction and airway remodeling by releasing histamine and tryptase (Louahed et al. 2000, 2001). These cells also release proinflammatory cytokines such as IL-4 and TNF-α, neurotrophins, chemokines, and growth factors in response to the stimulation by allergens via an IgE-dependent mechanism, involving FcϵRI receptors. These mediators are known to play vital roles in pathophysiology of asthma (Molet et al. 2001a, b; Einarsson et al. 1995).

5.3   Macrophages

Macrophages produced by the differentiation of blood monocytes are recruited to the airways during asthmatic conditions. These cells are activated by allergens via their FcϵRII receptors (Creticos et al. 1984; Shindo et al. 1994). These cells produce certain inflammatory proteins which are known to coordinate the inflammatory response in asthma. One of the proteins secreted by macrophages is IL-10, which attenuates the expression of Th1 cytokines and MHC class II antigens (Cowburn et al. 1998). Macrophages are also implicated in the regulation of JAK/STAT pathway. Furthermore, macrophages are major phagocytic cells of the body (Sampson et al. 1997; Nasser et al. 1996).

5.4   Dendritic Cells

These are the prime antigen-presenting cells of immune system in mammals. These cells initiate an immune response, mediated by T-lymphocytes and hence play an essential role in asthma development (Henderson 1994). The dendritic cells act as the most effective antigen-presenting cells in the respiratory tract of mammals. So, it is likely that these cells are essential in the instigation of allergen-induced responses in asthma (Hay et al. 1995; Wanner et al. 1996).

5.5   CD4+T Cells

5.5.1   Activation and Differentiation

Stimulation of T cell receptors with antigen, followed by presentation and processing of the antigen by professional antigen-presenting cells like dendritic cells marks the start of differentiation process of these cells. This antigenic activation of CD4+T cells induces a cascade of downstream signaling pathways leading to differentiation of these cells into specific effecter cells. These effecter cells are grouped into Th1, Th2, Th17, and regulatory T cell subsets, based on their cytokine signature. For CD4+T cells to differentiate into particular effecter cells, certain factors are necessary like the cytokine milieu of the microenvironment, type of antigen-presenting cell, and concentration of antigens (Ashkar et al. 2000).

Interleukin 12 and IFN-γ are the prime cytokines that commence the signaling cascade for the development of Th1 cells (Trinchieri et al. 2003). IL-12 is secreted in bulk amounts by antigen-presenting cells after their activation (Iwasaki and Medzhitov 2004; Trinchieri and Sher 2007). The IL-12, in turn, stimulates the production of IFN-γ by natural killer cells, thus enhancing Th1 cell differentiation. Several transcription factors are responsible for inducing Th1 cell differentiation like T-box transcription factor (T-bet), STAT1 and STAT4. However, T-bet is of prime importance which restrains the differentiation of Th2 and Th17 cells by impairing the function of GATA3 and RORγt (Afkarian et al. 2002). For differentiation of Th17 cells, the main signaling cytokines involved are TGF-β, IL-21, IL-6, and IL-23, and the prime transcription factors involved are retinoic acid receptor-related orphan (RORγt) and STAT3. However, RORγt is the master regulator of Th17 cell development and differentiation (Veldhoen et al. 2006; Bettelli et al. 2006). In case of regulatory T cells, TGF-β is the decisive cytokine accountable for the differentiation process and FOXP3 (forkhead box P3) is the chief lineage-specific transcription factor responsible for Treg cell differentiation (Chen et al. 2003; Li et al. 2007).

5.5.2 Th2 Cells and Asthma

Th2 cells are factually recognized by their cytokine signature. The cytokines released by these cells mainly include IL-4, IL-13, and IL-5. These cytokines primarily control the key events, which epitomize inflammatory asthmatic response, such as airway hyperresponsiveness, migration, and activation of inflammatory cells especially eosinophils to airways, mucus production, and IgE isotype switching. Th2 cells comprise the chief and most important cells involved in the pathobiology of asthma observed both in humans and mice studies (Mosmann et al. 1986). This correlation between Th2 cells and asthma has been confirmed in various murine models of AHR, which demonstrated Th2 cells as the most predominant cells to mediate AHR, airway eosinophilia, and goblet cell hyperplasia, after exposing mice to a suitable antigen through inhalation (Cohn and Ray 2000).

IL-4 is the critical Th2 cytokine responsible for Th2 cell differentiation which it does via a STAT6-dependent pathway. However, Th2 cell differentiation also occurs via a STAT6-independent pathway, involving STAT5 and IL-2 receptor signaling. This pathway involves IL-2-induced STAT5 activation causing epigenetic modifications of IL-4 at sites on DNA that are different to those modified by GATA3 (after activation by STAT6 via STAT6-dependent pathway), and this leads to IL-4 production by Th cells. However, Th2 differentiation via this pathway occurs with reduced efficacy. So, IL-4-induced STAT6-dependent pathway is the main route to Th2 cell differentiation from naive Th cells (Ho et al. 2009; Rayees et al. 2014). The significance of IL-4 and STAT6 in Th2 cell differentiation has been documented in various studies by generation of IL-4 or STAT6 target-specific deficient mice (Glimcher and Murphy 2000; Zhu et al. 2001).

Chapter 6
Transcription Factors Involved in Th2 Cell Differentiation

The chief transcription factors associated with Th2 lineage differentiation include STAT6 and GATA3. GATA3, which is upregulated by phosphorylated STAT6, instigates the transcription of downstream Th2-specific genes. It is the principal transcription factor for differentiating naive Th cells into Th2 cells (Glimcher and Murphy 2000; Zhu et al. 2001). However, GATA3 alone cannot modulate the transcription of these genes, but its collaboration with STAT6 is essential which was confirmed by several studies, where STAT6-deficient mice showed diminished IgE titers and Th2 differentiation was blocked (Horiuchi et al. 2011; Shimoda et al. 1996). Binding of IL-4 to its receptor (IL-4Rα) stimulates Jak1 and Jak3 (JAK family of receptor-associated Kinases) leading to the phosphorylation of STAT6, followed by its dimerization (homodimers). This modulated form of STAT6 translocates from cytoplasm to the nucleus, binding to its specific DNA sequences and activating transcription of cytokine-responsive genes especially GATA3 (Fig. 6.1) (Zhu et al. 2006). However, the source from which this initial IL-4 is produced that causes the phosphorylation of STAT6, and hence the activation of GATA3 in vivo is still unclear (Ho et al. 2009). GATA3 involvement in Th2 differentiation involves different mechanisms which include enhanced Th2 cytokine production, and inhibition of Th1 differentiation, presumably by modulating the expression of T-bet (Zhu et al. 2006; Ouyang et al. 2000). GATA3 activates the transcription of IL-5 and IL-13 genes by directly binding to their promoters (Zhou and Ouyang 2003). Moreover, this transcription factor also causes remodeling of chromatin structure and opening of IL-4 locus (Ouyang et al. 2000). GATA3 also downregulates STAT4 expression, another Th1-specific transcription factor (Usui et al. 2003). Further, the obligatory role of GATA3 for Th2 differentiation has been demonstrated in various in vivo studies performed on GATA3-deficient mice, where either the naive Th cells side tracked towards Th1 lineage (Zhu et al. 2004) or a significant interruption of Th2 differentiation was observed (Zhu et al. 2004, 2006; Pai et al. 2004).

Differentiation of Th2 cells can also occur independently of STAT6 through GATA3 expression, but with reduced efficiency. Studies have indicated that Notch

© The Author(s), under exclusive licence to Springer Nature Switzerland AG 2021   17
S. Rayees, I. Din, *Asthma: Pathophysiology, Herbal and Modern Therapeutic Interventions*, SpringerBriefs in Immunology,
https://doi.org/10.1007/978-3-030-70270-0_6

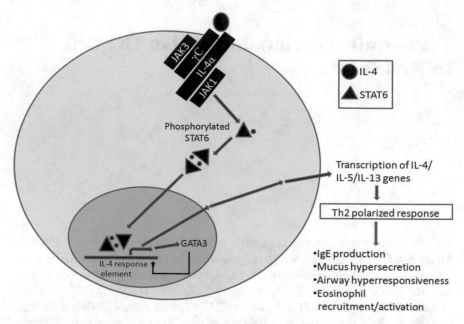

Fig. 6.1 Stimulation of CD4+T cell by IL-4, causing expression of Th2 cytokines via STAT6 phosphorylation and GATA3 expression. GATA3 is crucial for the transactivation of IL-4, IL-13, and IL-5 genes, which further leads the enhanced expression and production of other cytokines like IL-6, IL-9, and IL-10 and also other mediators like immunoglobin E, which also contribute to enhance the pathologies seen in asthma

signals (Notch proteins are a family of type I transmembrane proteins) regulate the transcription of GATA3 directly and enhance Th2 differentiation (Ho et al. 2009). Transcription of GATA3 is initiated from two promoters located ~10 kb apart (Fang et al. 2007). Both promoters transcribe GATA3 in CD4+ T cells; however, the distal promoter is mainly used in brain tissue. The distal promoter contains a consensus binding site for RBPJ and is responsible for GATA3 transcription by Notch signaling (Rayees et al. 2014; Ho et al. 2009).

Chapter 7
Current Asthma Treatments

Asthma is an intricate respiratory disorder with differences in its severity, natural history, and hence treatment response. These differences in intensities of various phenotypes like bronchial hyperresponsiveness, airway inflammation, mucus production, and airflow obstruction make asthma a heterogeneous disease. However, when asthmatic patients are studied in large groups, this definition of asthma of being a heterogeneous disease doesn't seem appropriate because all the patients studied will have two or more common features/phenotypes (Lotvall et al. 2011). Most often, there is a difference in intensities of airway eosinophilia among asthmatic patients. In fact, a subgroup of asthma patients does not have airway eosinophilia (non-eosinophilic subtype) although the size of this subgroup is uncertain. This heterogeneity in asthma phenotypes especially eosinophilia marks a significant effect on the efficacy of some current asthma drugs especially those that target eosinophilic asthma. The mainstay of current therapies for asthma includes inhaled corticosteroids, phosphodiesterase inhibitors, leukotriene modifiers, and β2-adrenoceptor agonists (Mcgrath et al. 2012; Gibeon and Menzies-Gow 2013).

7.1 Inhaled Corticosteroids

Corticosteroids possess strong anti-inflammatory and immunosuppressive properties. Inhaled corticosteroids (ICS) are known to improve lung dynamics more efficiently than bronchodilators (Barnes and Adcock 2003). They are considered as the most effective strategy for long-term asthma management. One of the first inhaled corticosteroids Beclomethasone was recommended for use four times a day and each puff of medicine from a metered-dose inhaler, marketed in the United States, produced only 42 μg of medicine (Barnes 1998a, b). Since then, more corticosteroid formulations have been developed, having more active medications which provide an optimum dosage per inhalation and are approved for usage once or twice a day (Suissa et al. 2000).

ICSs drug class help patients to achieve well-controlled asthma (Cochrane et al. 2000). ICS treatment has been shown to have significant impact on the reduction of morbidity and mortality correlated with asthma, when given in combination with alternative drugs including β2-agonists, which may offer symptomatic relief. Inhaled corticosteroids reduce inflammation of the airways; however, they are unable to infiltrate far into the airways (Rowe et al. 2001). They are usually effective once daily in patients receiving low doses. ICS therapy has significantly fewer side effects due to limited systemic absorption (Campbell 1999).

Studies are yet to establish whether ICS in a systematic intervention has a healing impact on the airways. Experiments performed in dogs showed that inhaled steroids influence the development of leukocyte progenitors in the bone marrow although it is not certain if this arises from the creation of any activating agent in the airways or from the regularly ingested fraction of inhaled steroids on the bone marrow (Woolley et al. 1994).

Implementing ICS early has a potential to improve lung dynamics more than treatment with bronchodilators. The logical corollary is that all patients should be administered inhaled steroids as soon as initial asthma is detected though this needs more long-term, regulated trials to confirm systematic implementation of this program. Since long-term experiments gave promising safety results, the usage of inhaled steroids grew tremendously. Pharmacotherapy guidelines are given with updated guidance for the usage of inhaled corticosteroids (Suissa and Ernst 2001). Inhaled corticosteroids are usually effective in patients receiving a single low dose daily. Once ingested and systemically consumed from the gastrointestinal tract, before they enter bloodstream circulation, most of the commercially accessible inhaled corticosteroids experience significant first-pass mitochondrial inactivation in the liver. In addition, since less than 20% of the dosage administered is typically concentrated in the airways, significant doses are eligible for systemic absorption via the mucosa of the respiratory tract (Selroos et al. 1995).

ICS is the most effective long-term asthma management treatment for reducing airway hyperresponsiveness (EPR-3 2007). Upon comparison with oral corticosteroid therapy, ICS therapy has significantly fewer side effects due to limited systemic absorption. ICS drugs have also seen important impacts on the reduction of morbidity and mortality correlated with asthma (Childhood Asthma Management Program Research Group 2000). A clinical study performed on patients who had experienced asthma death or near death showed that daily usage of low-dose ICSs was correlated with a decreased chance of death from asthma (Ernst et al. 1992). Several studies show that the long-term application of inhaled corticosteroids may reduce the gradual decline of lung function in patients with asthma (O'Byrne et al. 2009; Childhood Asthma Management Program Research Group 2000).

7.2   Adverse Effects

Dysphonia and candidiasis are the two major adverse effects of ICSc. Dysphonia has been reported to occur in about 50% of the patients undergoing high-dose therapy. The pharyngeal and laryngeal side effects include sore throat, medication coughing, low or loud speech, and candidiasis. Washing the mouth with a metered-dose inhaler after increasing treatment and utilizing a valved keeping chamber are two strategies which may reduce the risk of oral candidiasis (Dahl 2006; Roland et al. 2004). Systemic side effects include skin swelling, atrophy, and decreased mineral density of the bone. Systemic adverse effects are caused by the ingestion in GI tract from the swallowed portion of the medication as well as the portion stored in the lung. Appropriate medical care while prescribing ICS correctly will help to reduce such side effects (Agertoft and Pedersen 2000). Not all individuals gain similarly from the inhaled corticosteroids usage. For starters, active tobacco users are less likely to experience the same anti-asthmatic results as non-smoker ones. ICS performance is usually moderate as people do not expect direct gain from the procedure and may often have questions about adverse effects of steroids (Buhl 2006).

7.3   Leukotriene Antagonists

Leukotrienes are a family of lipid mediators which contribute to inflammation of the airways and smooth contraction of the muscles. Leukotrienes were first discovered by Samuelsson et al. in 1970s and are considered as a potential pharmacological target in lung pathological conditions (Drazen et al. 1999). Leukotriene antagonists or leukotriene modifiers (LTMs) include Montelukast, Zafirlukast, and Pranlukast. Montelukast is made available in chewable tablets and oral granules for small children. Leukotriene antagonists act by obstructing cysteinyl leukotriene receptor. Antagonists to leukotriene receptors can be issued as tablets once (Montelukast) or twice daily (Zafirlukast). Legal recommendations provide preventive medication for chronic asthma in moderate to serious conditions (Finnerty et al. 1992). Though LTMs tend to be inferior to ICS monotherapy for the management of chronic asthma, LTMs are very beneficial in patients with aspirin-exacerbated respiratory disorder. Bronchodilatation happens within hours of first dosage and is highest during the first few days of administration. In United Kingdom, leukotriene antagonists are officially approved for use in patients that are symptomatic following inhalation of corticosteroids (Lam et al. 1988).

Several trials found that the usage of the recommended doses of leukotriene antagonists enhanced lung capacity, provided better ratings on asthma-related quality of life questionnaires and less asthma attacks (Robinson et al. 2001; Balzano et al. 2002; Bjermer and Diamant 2002). In fact, patients with severe asthma, who smoke or who have related vulnerability to aspirin can benefit from leukotriene antagonist medication. Montelukast and Zafirlukast are both powerful cysteinyl leukotriene receptor antagonists capable of efficiently inhibiting exercise-induced bronchoconstriction and the early and late reaction to allergens inhaled. Leukotriene antogonists are an effective treatment for a moderate chronic asthma (Holgate et al. 1996).

7.4 Adverse Effects

In some patients Zileuton, another Leukotriene antagonists is known to cause reversible chemical hepatitis. In children taking Montelukast, few post-marketing studies reported depression (Israel et al. 1996). Evidence of the development of Churg–Strauss syndrome in patients on a leukotriene receptor antagonist has also been reported (Jamaleddine et al. 2002). But in general, the antagonists of the leukotriene receptors were regarded as virtually free of side effects. The medication-induced mood and behavioral changes in patients could be possible due to LTMs.

7.5 Short-Acting Beta2 Agonists

Short-acting beta2 agonists (SABAs) are used to treat moderate and sporadic asthma. The SABAs may be classified into fast- and long-acting pharmaceutical items as they have an impact starting within 5 min, with a median effect of 30–60 min or 4–6 h (Walters et al. 2003). Inhaled SABAs such as Salbutamol and Terbutaline are powerful bronchodilators and are recommended for individuals suffering from symptomatic asthma. SABAs resolve acute asthma symptoms and may initially be used for the first hour every 15–20 min during acute asthma. They are also useful when given before the start of exercise to avoid exercise-induced asthmatic symptoms. SABAs have no effect on inflammation of the airways or sustained benefit. There is no effective anti-inflammatory activity in them. The determination on which different β-agonists such as Salbutamol, Levalbuterol, and Pirbuterol are used primarily depends on the patient or practitioner expense and choice (Vezina et al. 2014).

The SABAs are hydrophilic and their metabolism is quick. SABAs are developed to treat moderate, sporadic asthma, with marginal action of α-receptors but strong responsiveness to β2-receptors. The medication is given both orally and nebulized by metered-dose inhaler (Griffiths and Ducharme 2013).

7.6   Adverse Effects

SABAs are sympathomimetic agents and are sometimes used in combination with anti-inflammatory medications, as they are ineffective on inflammation alone. The large doses are correlated with elevated risk of mortality and morbidity in asthma (Vezina et al. 2014). The short-acting β2-agonists become dangerous even when used at large doses, so the likelihood of mortality even rises dramatically if more than one canister is used every month. High doses can cause tremor, palpitations, and tachycardia. Short-acting β2-agonists in asthmatic airways may not minimize inflammation and there is some indication that inflammation may increase. Short-acting β-agonists to be ingested are discouraged in either tablet or liquid form because they take longer to start operation, are less active, and are correlated with more severe side effects than short-acting β-agonists inhaled (Sestini et al. 2002).

7.7   Long-Acting Beta2 Agonists

Long-acting β2-agonists are a recommended choice especially in situations where inhaled corticosteroids do not function effectively. Prominent ones among this category include Salmeterol and Formoterol though their binding efficiency for β2-receptors varies from partial to full, respectively (Nelson et al. 2006; Dhand et al. 2012; Cazzola and Donner 2000). Highly selective β2-agonists such as LABAs demonstrate greater regulation of the effects and respiratory activity relative to Albuterol (Fuso et al. 2013). Several studies have demonstrated increased regulation of effects and greater lung function by using LABAs though high doses can cause palpitations and tachycardia. They are also useful to avoid exercise-induced asthmatic symptoms (Nelson et al. 2006; Dhand et al. 2012). Yet LABA monotherapy is not recommended for long-term control of asthma (Ram et al. 2005). In most of such cases, inhaled corticosteroid drug therapy is used to treat people with mild to serious chronic asthma. Probably due to their lack of anti-inflammatory function prevents their use as first-line drugs. So, the current recommendation is that they should be prescribed alongside the ICSs (Ram et al. 2005; Cazzola and Donner 2000).

In a double-blind clinical trial, it was found that using Salmeterol and moderate doses of Fluticasone, an ICS, demonstrated a significant benefit to patients as compared to ICS alone. This study was performed in patients older than 12 years of age (Masoli et al. 2005). Patients had significantly heterogeneous reactions to inhaled Salmeterol, some have drastically improved symptoms, while some, particularly those with more extreme asthma, did not get any benefits. It was found that regular use of Salmeterol improved lung function and masked lung inflammation. However, the inflammation was delayed and not suppressed. So, this may be beneficial to provide a fixed-combination inhaler comprising long-acting β2-agonist and steroid. It could improve compliance and ensure patient receives a long-acting β2-agonist-inhaled steroid.

7.8   Adverse Effects

Compared with short-acting agents, LABSs show similar but less pronounced pharmacologically predicted side effects. High doses have been reported to sometimes cause palpitations, vomiting, and tachycardia. Formoterol and Salmeterol were reported to cause some dose-dependent mild side effects related to cardiovascular system and metabolism such as fluctuations in heart rate, blood pressure, and fluctuations in plasma glucose levels. They are also known to augment inhaled corticosteroid-induced side effects (Guhan et al. 2000; Kips and Pauwels 2001).

7.9   Phosphodiesterase Inhibitors and Methylxanthines

Phosphodiesterase (PDE) 4 inhibitors are used as an alternative treatment for asthma and chronic obstructive pulmonary disease (COPD) because of their wide spectrum anti-inflammatory effects. These drugs inhibit the functioning of T cells, mast cells, airway smooth muscles, and eosinophils. However, PDE-4 inhibitors are not able to attenuate major pathological features of asthma like airway hyperresponsiveness, bronchoconstriction, etc. to a significant extent. Commonly used PDE4 inhibitors in asthma and COPD include Roflumilast and Cilomilast (Beghe et al. 2013; Franciosi et al. 2013; Bousquet et al. 2006; Fan Chung 2006).

Methylxanthines have been used since the 1920s to combat asthma and other pulmonary disorders but soon dropped out following the introduction of β2-agonists (Lam and Newhouse 1990). Theophylline is an effective bronchodilator and certainly the most powerful methylxanthine which acts by increasing cyclic adenosine monophosphate (cAMP) levels due to its strong phosphodiesterase inhibitory action. For theophylline to be effective, higher serum concentrations have to be achieved. It is often believed that side effects are generally associated with the drug's plasma concentration (Barnes 2003; Barnes and Pauwels 1994; Barr et al. 2003). However, at lower concentrations it is reported to mask inflammation by suppressing the release of mediators from mast cells (Barnes 2013; Barr et al. 2003).

Theophylline is recommended in conjunction with ICSs. While theophylline once was a widely used agent for asthma treatment, its usage has been restricted by the comparatively high incidence of adverse effects. The plasma rates have to be carefully controlled owing to recurrent side effects. Adverse effects may be ameliorated in certain situations by gradually increasing the dosage depending on blood rates and clinical side effects. The most severe side effects include headache, diarrhea, and gastric reflux (Barnes and Pauwels 1994).

An essential benefit of theophylline is that it can be ingested by mouth as a treatment for gradual release once or twice a day. The use of this oral treatment can enhance patient consistency as contrasted with the use of inhaled medicine. Theophylline is especially helpful in the care of those with more serious asthma and is also effective for asthma at night (Barnes and Pauwels 1994; Barnes 2013; Barr et al. 2003).

7.10   Adverse Effects

Theophylline has comparatively fewer side effects. Cardiac arrhythmias and epilepsy are more severe ones but are less frequent. Side effects are generally associated with the drug's plasma concentration. Its large doses are correlated with elevated risk of mortality and morbidity with asthma. General side effects include headache, diarrhea, palpitations, and gastric reflux (Barnes and Pauwels 1994; Barnes 2013, Barr et al. 2003).

7.11   Anticholinergics (Anti-muscarinic)

For patients with COPD, anticholinergic bronchodilators are the first-line treatment and are considered as better than any other medication (Cazzola and Donner 2000; Matera et al. 2014). They are less successful in patients with asthma than β2-agonists as they block only the cholinergic portion of bronchoconstriction. The β2-agonists reverse all bronchoconstrictors, including the direct effects of inflammatory mediators such as histamine, leukotriene D4, and prostaglandin D2 (Rodrigo and Rodrigo 2002). Anticholinergic agents function by linking muscarinic receptors to a smooth muscle on the airway. Acetylcholine causes bronchoconstriction after its release from parasympathetic nerves followed by its binding to M3 muscarinic receptors. Anticholinergic bronchodilators cannot be used in the extreme event of an individual with resistance to both β-agonist bronchodilators and whether to manage serious asthma attacks. Nonetheless, in asthma, eosinophilic recruitment results in decreased release of acetylcholine owing to the development of a large specific protein, an antagonist of the M2 receptor (Choby and Lee 2015; Rodrigo and Rodrigo 2002).

The medication is distributed both orally and in nebulized manner. They have no effect on inflammation of the airways or sustained benefit. Nebulized ipratropium bromide is almost as effective in managing acute extreme asthma as nebulized beta2 agonists (Westby et al. 2004).

7.12   Adverse Effects

This class of drugs does not have a significant impact on the reduction of morbidity and mortality correlated with asthma. Their half-life is also a major issue. Ipratropium often needs dosage per 6 h in order to preserve efficacy which restricts its clinical usage. Anticholinergic bronchodilators cannot be used to manage serious asthma attacks or in the patients with resistance to both β-agonist bronchodilators. (Westby et al. 2004; Choby and Lee 2015; Rodrigo and Rodrigo 2002).

Chapter 8
Cytokine-Based Therapies

The past decade has given a better understanding of pathobiology of asthma and its cellular and molecular mechanisms. The awareness of various inflammatory and immune events has unraveled the crucial role of Th2 cells and the cytokines they release, in all forms of asthma. They are regarded as prime drivers of mild to moderate asthma. Thus, they have significant therapeutic implications in asthma. These cytokines along with other chemokines have been targeted in different ways, involving the use of biologics like humanized blocking mAb to their receptors, elimination of cytokines or chemokines by their binding to soluble receptors, antibodies that block specific inflammatory agents like IgE or using small molecule receptor antagonists (Fig. 8.1) (Walsh 2013a, b, 2011; Gallelli et al. 2013). The biologics or biologicals include protein-based therapies like antibodies (e.g., Lebrikizumab), recombinant protein-based receptor antagonists (e.g., Pitrakinra) and soluble receptors (e.g., Etanercept) (Cook and Bochner 2010). Targeting these cytokines can have a significant impact on pathophysiological features of asthma, which occur as a result of their function. Currently, various anti-asthma biologics are under investigation, including molecules targeting Interleukin-4, Interleukin-9, Interleukin-5, and Interleukin-13.

8.1 Anti-IL-9 Therapy

IL-9 facilitates the mounting of airway inflammation, mucus secretion, differentiation, and development of mast cells in asthma and in large by augmenting the expression of other Th2 cytokines (Oh et al. 2011). Blocking IL-9 promotes growth of T cell (Goswami and Kaplan 2011). Within human lungs, IL-9 can control the phenotype of several cell types especially mast cells. Several trials found that the usage of the recommended dosage enhanced lung capacity, better ratings on asthma-related quality of life questionnaires and less asthma attacks (Levitt et al. 1999; Goswami and Kaplan 2011).

© The Author(s), under exclusive licence to Springer Nature Switzerland AG 2021 27
S. Rayees, I. Din, *Asthma: Pathophysiology, Herbal and Modern Therapeutic Interventions*, SpringerBriefs in Immunology,
https://doi.org/10.1007/978-3-030-70270-0_8

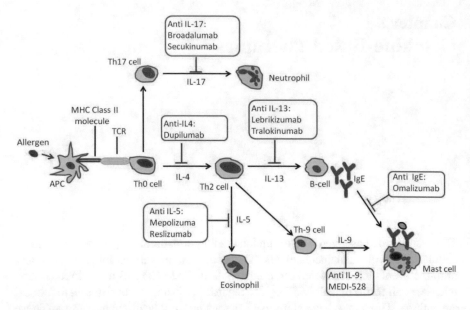

Fig. 8.1 Cytokine-based therapies: Potential cytokine targets and the monoclonal antibodies developed against them are shown, as a representative diagram. The detailed explanation is given in the text

Blocking IL-9 acts by decreasing the inflammation of the airways caused by allergens and hyperresponsiveness of the airways. This displays strong anti-inflammatory function and therefore may be useful in clinical setting. Few human-ized monoclonal antibodies against IL-9 were tried on safe volunteers with no complications and the efficacy was the limiting factor. MEDI-528 is an IL-9-specific humanized monoclonal antibody for symptomatic moderate to severe asthma that has shown promising results in preclinical studies, however lacks intensity in treating human subjects. It was found that 50 mg of MEDI-528, given twice a week subcutaneously, is capable of exercising a protective effect against bronchoconstriction (Parker et al. 2011a, b). Adverse effects may be ameliorated in certain situations by gradually increasing the dosage depending on clinical side effects. There was no substantial improvement observed in the asthma management questionnaire's ratings, exacerbations, pulmonary capacity, or quality of life. However, there was no safety concern noted (Oh et al. 2013; Humbles et al. 2010).

8.2   Anti-IL-13 Therapy

IL-13 facilitates the growth of goblet cells and their secretion in the airway lumen, impairs the smooth reaction of the airway to beta-adrenergic agonists, and enhances the production of submucosal fibroblasts and collagen deposition (Padilla et al.

2005). IL-13 is regarded as the central mediator of allergic asthma (Wills-Karp et al. 1998). It plays a key role in asthma pathology and has a direct impact on the development of airway hyperresponsiveness, mucus gland hyperplasia, chronic elevation of serum IgE, and eosinophilic airway inflammation, rendering IL-13 as a potent therapeutic target in asthma. GSK 679586, Tralokinumab, Anrukinzumab, and Lebrikizumab are IL-13-specific humanized monoclonal antibodies developed till date which had a significant impact on asthma reduction either in preclinical or clinical setting; Hacha et al. 2012; Hambly and Nair 2014; Quirce et al. 2014). GSK 679586, a humanized IgG1 anti-IL-13 monoclonal antibody now under Phase II study, showed a dose-dependent pharmacological effect in a randomized, placebo-controlled Phase I dose escalation trial (Hodsman et al. 2013). However, it was not able to improve the lung function or exacerbations in severe asthma patients. Though there were no serious safety concerns noted, it could not demonstrate an overall control on severe asthma (De Boever et al. 2014).

Tralokinumab is another humanized anti-IL-13 antibody which is in clinical development. In a randomized, double-blind, placebo-controlled, Phase IIa trial performed on 194 asthmatic subjects, tralokinumab was able to resolve overall lung function, but no improvement in Asthma Control Questionnaire score was noted. However, no safety concerns were noted as well (Piper et al. 2013). In a placebo-controlled, multicenter and double-blind clinical trial Tralokinumab did not demonstrate any resolving effect and did not significantly affect eosinophilic inflammation/counts evaluated from blood or sputum. However, a significant reduction was noted in serum IgE levels and also fractional exhaled nitric oxide. This trial questions the role of IL-13 in pulmonary eosinophilic inflammation in asthma (Russell et al. 2018).

8.3   Anti-IL-4 Therapy

IL-4 is essential in asthma development because of its imperative role in Th2 cell differentiation via STAT6 phosphorylation. It induces IgE isotype switching and helps in the recruitment and activation of eosinophils, basophils, T-lymphocytes, and monocytes to the sites of inflammation (Ramalingam et al. 2008; Sokol et al. 2008).

Blocking the production of IL-4 or inhibiting its functioning using a biological agent has been documented to have reflective effects on asthma phenotypes, mostly in animal models. Dupilumab is a monoclonal antibody which blocks IL-4 function by binding to its receptors, i.e., IL-4 and/or IL-13 (Wechsler 2013). In a double-blind, placebo-controlled trial of Phase II, dupilumab was able to reduce asthma exacerbations and FEV also. It however had no impact on eosinophil reduction (Vatrella et al. 2014). Dupilumab has a favorable safety record, with mild side effects involving an injection site reaction and intermittent blood eosinophilia. Dupilumab was licensed by the FDA for the diagnosis of atopic dermatitis and has recently been licensed for asthma care (Vatrella et al. 2014). Another humanized monoclonal antibody Pascolizumab was evaluated for its safety and efficacy by Hart et al. With a long half-life, it did well in preclinical setting. No toxic effects were noted though slight tissue accumulation was found in chronic dosages. It did not produce desired effects on IgE reduction in clinical trials which were aborted later (Bagnasco et al. 2016; Cook and Bochner 2010). However, most of the IL-4 monoclonal antibodies failed to justify their efficacy in human subjects, e.g., Altrakincept, a soluble, non-immunogenic recombinant human IL-4 receptor failed in Phase III trials to validate its efficacy as an anti-asthma candidate. Further, Pascolizumab, clinical trials Pascolizumab were also aborted due to efficacy issues (Bagnasco et al. 2016; Santini et al. 2017).

8.4 Anti-IL-17 Therapy

Th17 cells are known to secrete cytokine IL-17 which marks their contribution in asthmatic inflammation. The IL-17 molecule has multiple subtypes, i.e., type A, B, C, E, and F. It plays a significant part in controlling the airway movement of neutrophils and is active in fibrosis and airway remodeling (Nembrini et al. 2009; Agache et al. 2010). It was found that IL-17 significantly upregulated in asthmatic lungs as seen from the bronchial biopsies collected asthmatic subjects. The blood and airway levels of IL-17 have also been noticed to increase in animal studies of asthma. Several significant studies done till now have indicated IL-17 as a potential drug goal in asthma (Agache et al. 2010; Bullens et al. 2006; Molet et al. 2001a, b; Chesné et al. 2014).

Brodalumab and secukinumab are two monoclonal antibodies developed to target IL-17 in asthma. Brodalumab is human unique monoclonal IL-17-specific antibody recommended for patients where asthma is not sufficiently controlled by inhaled corticosteroids and/or LABAs. It inhibits the biological function of IL-17A, IL-17F, IL-17A/F heterodimer, and IL-17E (IL-25) and is under clinical trials (Busse et al. 2013). Secukinumab is an IgG1κ monoclonal antibody targeted at IL-17A. Secukinumab has completed its Phase II of clinical trials for uncontrolled asthma. The results have been significantly good when compared to placebo control as per the Asthma Quality Control score. Secukinumab has shown significant results in psoriasis and rheumatoid arthritis also. However, more scientific trials therefore need to confirm their medicinal efficacy, tolerability, and safety (McCracken et al. 2016).

8.5 Anti-IgE Therapy

Allergic asthma accounts for around 70 per cent of asthma and IgE is central in the allergic asthma inflammatory cascade (Vernon et al. 2012). In allergic asthma, IgE rates are usually elevated and are often correlated with allergic symptoms (Soler et al. 2001). IgE is an essential mediator involved in various allergic disorders, including asthma. It plays a central role in the development and degranulation of mast cells and basophiles as a result of its binding to high affinity (FcεRI) and low affinity (FcεRII) receptors on these cells. Further, it is associated with antigen recognition processes by dendritic cells and maintenance of inflammation in the airways (Barnes 2010; Walsh 2011). The decreased association between IgE and FcÿRI inhibits degranulation caused by allergens to dramatically reduce the effects caused by IgE. Anti-IgE antibodies on the surface of mast cells and B-lymphocytes may bind to serous IgE and membranous IgE (Vernon et al. 2012; Soler et al. 2001).

Omalizumab is a humanized antibody that has been accepted as an antidote to therapeutic steroids for managing extreme allergic asthma. It has been widely approved for managing allergic asthma for over 15 years (Normansell et al. 2014). Omalizumab results in a rapid decrease in circulating IgE levels as a result of its binding to high affinity FcεRI receptors. Omalizumab has been reported to inhibit both early and the late phase asthmatic reactions and diminish circulating IgE levels significantly. It has been found beneficial in patients with moderate to severe forms of asthma or patients whose asthma was poorly controlled by inhaled corticosteroids. In fact, Omalizumab usage causes a reduction in the intake of corticosteroids and β2-agonists (Ayres et al. 2004; Bousquet et al. 2005). The findings from these trials describe Omalizumab treatment as an effective and reliable therapy for asthma whether mild, moderate, or severe. However, Omalizumab treatment is associated with some adverse effects like anaphylaxis and systemic vasculitis (Holgate 2014; Humbert et al. 2014).

Ligelizumab (QGE-031; Novartis), MEDI-4212 (AstraZeneca), and Quilizumab (anti-M1 prime mAb; Roche) are other monoclonal anti-IgE antibodies (Gasser et al. 2020; Harris et al. 2016).

8.6 Anti-IL-5 Therapy

IL-5 is of prime importance in eosinophil survival, maturation, and differentiation process. It is the main cytokine that serves to attract, activate, and withstand eosinophils. It augments immunoglobulin secretion through its binding to IL-5 receptor and stimulates the growth of B cells (Ichinose and Barnes 2004; Walsh 2011). Two humanized IL-5-specific monoclonal antibodies Mepolizumab and Reslizumab were developed. Mepolizumab was developed for the treatment of eosinophilc asthma, atopic dermatitis, eosinophlic esophagitis, and hypereosinophilic syndrome and its Phase III clinical trials were finished in 2014. Mepolizumab has been shown to have a protection profile comparable to placebo. Mepolizumab is only licensed for an eosinophilic phenotype for adults aged 12 years and over, who have serious asthma. It is well tolerated, which however may sometimes contribute to symptoms of hypersensitivity, including anaphylaxis or urticaria (Stein et al. 2006; Ortega et al. 2014). Similarly, Reslizumab was developed for eosinophil-mediated airway inflammation. Reslizumab is undergoing Phase III clinical trials. Reslizumab is licensed for patients 18 years of age and older with serious eosinophilic asthma, receiving corticosteroid treatment. It however is not pre-scribed for acute bronchospasm or even other eosinophilic situations other than asthmatic (Hom and Pisano 2017; Walsh 2013a, b). However, over a course of 10–12 weeks, both of these monoclonal antibodies have been demonstrated to have no effects on either airway hyperresponsiveness or any measures of asthma outcome (O'byrne 2013; Walsh 2009).

8.7   Anti-TSLP

Thymic stromal lymphopoietin (TSLP) belongs to the cytokine family IL-2 (Ziegler et al. 2013). The bronchial epithelium releases it in reaction to tissue activation or harm incurred by mechanical injury and pro-inflammatory mediators (TNF-a, IL-1-a, IL-4, IL-5), allergens, and proteases. Upon entry, TSLP contributes to acti-vation or suppression of a variety of immune cells (Morshed et al. 2012; Ziegler et al. 2013). TSLP is known to modulate the dendritic cell activity, which then facilitates the maturation of naive CD4 T cells into CD4 Th2 cells. Further, TSLP is reported to augment the secretion of cytokines from several immune cells such as mast cells, natural killer T cells, and eosinophils and also aids in the functioning of a subset of basophils (Castillo et al. 2018; Morshed et al. 2012; Ziegler et al. 2013). Tezepelumab is a humanized antibody and very specific to TSLP which acts as an antagonist of TSLP receptor. This antibody showed significant improvement in eosinophil reduction in both early and late phases of asthma development. Further, there was a significant decrease in fractional exhaled nitric oxide. In total, tezepe-lumab had a broad activity (Gauvreau et al. 2014; Corren 2019).

Chapter 9
Transcription Factor Modulators

Various components of the inflammatory cascade like cytokines, chemokines, adhesion molecules, and receptors are highly expressed in asthma. These proteins are not generally expressed in normal cells. This augmentation in their expression during the inflammatory process of asthma is due to the enhanced gene transcription induced in a cell-specific manner. This modulation in the expression of these genes is regulated by specific proteins called transcription factors which execute this process by binding to the specific DNA segments. So, transcription factors play a crucial role in asthma pathogenesis. Many transcription factors like STAT6, GATA3, NF-κB, and AP-1 are of prime importance in differentiating Th2 cells and therefore represent significant targets for asthma therapy. Various small molecule inhibitors or biologics that work via transcription factors are under investigation and several others (such as glucocorticoids) are already in clinical use (Caramori et al. 2008, 2004; Barnes 2006; Rayees et al. 2014).

9.1 STAT6

STAT6 is an important transcription factor that plays a crucial role in asthma pathobiology. It is mainly involved in Th2 cell differentiation process. STAT6 is principally stimulated via IL-4 signaling and is required for induction of airway hyperresponsiveness, mucus hypersecretion, and eosinophilia in murine models of asthma (Hoshino et al. 2004). It was observed that Th2 cytokines are not expressed in the lungs of STAT6-knockout mice and hence augmentation of the inflammatory process is attenuated after allergen challenge (Mathew et al. 2001). In addition, these animals didn't develop goblet cell metaplasia after IL-13 instillation. However, this process can be reversed by expression of the STAT6 transgene in such animals (Wills-Karp and Karp 2004). STAT6 is expressed only within bronchial epithelium and infiltrating cells of bronchial mucosa (Christodoulopoulos et al. 2001; Ghaffar

© The Author(s), under exclusive licence to Springer Nature Switzerland AG 2021
S. Rayees, I. Din, *Asthma: Pathophysiology, Herbal and Modern Therapeutic Interventions*, SpringerBriefs in Immunology,
https://doi.org/10.1007/978-3-030-70270-0_9

et al. 2000; Peng et al. 2004). Furthermore, a selective antisense RNA against STAT6 has been able to diminish the eotaxin release from human airway smooth muscle after stimulation with IL-13 or IL-4. So, STAT6 can prove to be a potential target for asthma therapy in future. Several approaches, involving small molecule inhibitors, dominant negative peptides, and antisense therapy directed at blocking STAT6 functioning, are under investigation, e.g., YM-341619 hydrochloride, AS1517499, and niflumic acid (Peng et al. 2004; Chiba et al. 2009; Oh et al. 2010; Nagashima et al. 2007).

9.2 GATA3

GATA3 is regarded as the master regulator of Th2 cell differentiation. Expression of GATA3 has been found to be significantly above normal level in T cells and bron-chial biopsies of asthmatic subjects (Caramori et al. 2004; Erpenbeck et al. 2003, 2006). GATA3 plays a crucial role in airway hyperresponsiveness, mucus hyperse-cretion, and development of airway eosinophilia in murine models of asthma (Kiwamoto et al. 2006). Furthermore, a blockade of GATA3 expression using anti-sense RNA in human T cells or an inhibition in its activity in these cells by a domi-nant negative mutant of GATA3, significantly attenuates various asthma phenotypes (Ouyang et al. 1998). So, targeting GATA3 by either using a small molecule inhibi-tor that could inhibit its functioning or by local delivery of GATA3 antisense may be novel approaches for asthma therapeutics. Furthermore, this approach has the advantage of suppressing the expression of other Th2 cytokines, especially IL-4, IL-13, and IL-5, whose expression is under direct control of GATA3 (Rayees et al. 2014; Finotto et al. 2001).

9.3 NF-κB and AP-1

These two transcription factors coordinate and exaggerate inflammatory processes in asthma. They are overexpressed in airway epithelial cells of asthmatic patients. Furthermore, NF-κB and AP-1 are important for Th2 cells to execute their functions in vitro (Barnes 2006; Caramori et al. 2004). Among several approaches to inhibit NF-κB, inhibition of IKKβ (inhibitor of nuclear factor kappa-B kinase subunit beta) by a small molecule inhibitor seems to be more promising. However, small peptide inhibitors of IKKβ/IKKγ association are an alternative strategy (Caramori et al. 2008). PNRI-299 is a small molecule inhibitor of AP-1. It reduces eosinophil infil-tration to airways, goblet cell hyperplasia, and IL-4 levels in a mouse model of asthma. It also causes a notable decrease in other Th2 cytokines and IgE levels. This compound is under clinical trials (Nguyen et al. 2003; Desmet et al. 2005).

Chapter 10
Preclinical Mouse Models of Asthma Used to Evaluate Drug Efficacy and Properties and Associated Drawbacks

The accurate biochemical processes and cellular events that occur in chronic airway inflammation in asthma are yet to be explored properly. Although the human asthmatics are the best subjects to investigate and explore these pathways and identify novel therapeutic targets, ethical issues are averse in performing experimental and molecular studies on human subjects.

Animal models, especially murine and guinea pig models, provide an alternative for assessment for mechanistic and biochemical understanding of asthma. However, due to the complexity and heterogeneity of this disease, it is dubious to employ a single model that will mimic all the characteristics of human disease whether morphological or functional. So using different animal models to investigate specific features of the disease is a viable alternative. The mouse is the most commonly used rodent, due to the easy availability of its transgenic varieties that mimic most of the human asthma features like airway inflammation, mucus secretion, cytokine and IgE secretion, and airway hyperreactivity (Nials and Uddin 2008; Kumar et al. 2008).

A variety of acute and chronic murine models, involving different sensitization and challenge procedures, have been developed for either therapeutic understanding of small molecule inhibitors, antisense or monoclonal antibodies, or elucidating the mechanisms associated with pathobiology of asthma. Acute murine models of asthma are widely employed in preclinical studies as they have been reported to reproduce various key features of human asthma including increased IgE and Th2 secretion, airway inflammation, AHR, and goblet cell hyperplasia (Kumar et al. 2008; Kumar and Foster 2002). Chronic murine models of asthma are employed to address specific issues related to acute models and more importantly used to elucidate the therapeutic effects of novel therapies after chronic exposure which involve an increased number of allergen challenges (Lloyd 2007).

The nature of a mouse model whether acute or chronic is mainly influenced by mouse strain employed for the study, type of allergen, and the sensitization and challenge procedures (Kumar et al. 2008; Zosky and Sly 2007). BALB/c is, however, the most commonly used strain as they optimally mimic Th2-based immunological response. However, other strains like A/J and C57BL/6 have also been used successfully in various studies involving allergen challenge (Boyce and Austen 2005; Kumar et al. 2008). Although diverse sensitization and challenge procedures are being used, the basic procedure remains the same involving systemic administration or different doses of allergen along with an adjuvant, mostly aluminum hydroxide which has been documented to enhance the development of Th2 response when administered with an allergen (Blyth et al. 1996; Boyce and Austen 2005). Different allergens along with an adjuvant are used for asthma induction like house dust mite, cockroach extracts, and ovalbumin (OVA). However, ovalbumin obtained from chicken egg is most commonly employed allergen. It induces vigorous airway inflammation, AHR, and pathological features in laboratory rodents mimicking human asthma. The allergen may be given through different routes like inhalation (aerosol), intranasal, or intratracheal, depending on the type of study (Johnson et al. 2004; Sarpong et al. 2003).

The major drawback associated with animal models is reproducibility of results in human subjects. Various drugs and therapies failed in clinical trials because several of these drugs/ therapies do not work with same the efficiency in human asthmatic subjects, which has been a topic of considerable debate (Holmes et al. 2011; Shin et al. 2009; Taube et al. 2004). Various in vivo studies depicted the importance of several cytokines and other inflammatory mediators in regulating asthma like IL-6, TNF-alpha, eotaxin, etc. which later lead to the development of several therapies against them but these therapies which appeared to have efficacy through these pathways in animal models have not been successful in human clinical trials (Walsh 2011; Walsh 2013a, b). However, presently there is no efficient alternative that mimics the human asthma to this extent as animal models do. These models cannot be considered a surrogate for human asthma but rather should be seen as an important opportunity to generate and test hypotheses in simple and controlled systems.

Chapter 11
Herbal Treatments of Asthma

The treatment of asthma by medicinal plants has been a long tradition in several countries (Saganuwan 2010). The medicinal plant must have some essential innate attributes for successful management of asthma. It must possess anti-histaminic and anti-inflammatory properties and must act as a smooth muscle relaxant. Herbs are the most important category of medicinal plants that are useful in successful supervision of asthma (Bielory and Lupoli 1999).

There is no iota of doubt towards herbal medicine being a successful and effective treatment for asthma. In fact, herbal medicine forms the core ingredient of the currently popular class of drugs (such as methylxanthines, cromones anticholinergics) for asthma (Lee 2000). Several researchers have suggested herbal medicine to play the role of complementary medicine for successful treatment of asthma (Graham and Blaiss 2000).

Traditional Indian herbal (ayurvedic) medicine for asthma involves four important medicinal herbs. A small herb called *Picrorhiza kurroa (P. kurroa)* has been the oldest and primitive herb for treating bronchitis and asthma (Sehgal et al. 2013). In Siddha system of southern India, the dry powder of *Solanum xanthocarpum and Solanum trilobatum* reportedly cure all the ailments of respiratory system (Govindan et al. 1999). Salai guggal, a gum resin, obtained from *Boswellia serrata (B serrata)* is composed of boswellic acids that stop the biosynthesis of leukotriene, which in turn improves respiratory performance (Ammon 2006). An indigenous plant of India, *Tylophora indica (T. indica)* has a long history in treating several patients suffering from asthma and respiratory ailments (Rani et al. 2012).

In line with the Indian herbal medicine, the traditional Japanese (Kampo) herbal medicine for asthma comprises of *Tsumura saiboku* (TJ-96). On the global level, TJ-96 stands among the most well-studied herbal medicine for asthma management (Urata et al. 2002). Based on these conclusions, we undertook to review all available herbal treatments of asthma.

Chapter 12
Traditional Plants with Anti-asthmatic Potential

Aerva lanata L. is a medicinal plant related to the Amaranthaceae family. It is mainly seen growing along roadside and most commonly identified by its small flowers with woollen texture. This plant is usually used to manage asthma (Kumar et al. 2009).

Ageratum conyzoides L. is a medicinal plant related to Asteraceae family, central to tropical America but its distribution is largely in tropical and subtropical areas of the world. This medicinal plant has a long list of reports for treating asthma patients all over the globe (Achola and Munenge 1998). *Argemone Mexicana* is a medicinal plant most commonly located in agricultural fields of India and known to possess anti-stress and anti-allergic properties. The leaves, stems, and roots of this plant have been customarily used to manage asthma and respiratory illnesses (Bhalke and Gosavi 2009).

Crinum glaucum and *Asystasia gangetica*, popularly named as creeping foxglove are medicinal plants common in the Nigerian region of Africa continent. Both have been successful to alleviate all the symptoms of bronchial asthma and cough relief (Okpo et al. 2001; Akah et al. 2003).

Acorus calamus proves extremely successful to completely eliminate the harmful effects of asthma in all age groups by discarding phlegm and catarrhal substances from trachea and tubes of the bronchia (Saxena and Saxena 2014).

Adhatoda vasica (L.) Nees (Acanthaceae), a small evergreen shrub commonly known as Malabar nut tree is used in indigenous system of medicine in India to treat various ailments. This plant is found in most parts of India up to an altitude of 1300 m. It is also found in Pakistan, Myanmar, Nepal, and Germany. The leaves of this plant are used to treat various respiratory disorders like asthma, cough, and bronchial obstruction (Gupta and Prajapati 2010; Hooper 1888; Kaur et al. 2013). *Adhatoda vasica* has been included in WHO manual *"The Use of Traditional Medicine in Primary Health Care"* for treating cough, asthma, and bleeding piles. This manual aims to facilitate the health workers of South-East Asia about the therapeutic uses of surrounding flora (Organization WH 1990).

© The Author(s), under exclusive licence to Springer Nature Switzerland AG 2021   39
S. Rayees, I. Din, *Asthma: Pathophysiology, Herbal and Modern Therapeutic Interventions*, SpringerBriefs in Immunology,
https://doi.org/10.1007/978-3-030-70270-0_12

In India, the leaf extract of *Aegle marmelos*, commonly called as Golden apple or the Bael fruit has been used to manage asthma, chest congestion, and several other chest-related ailments (Sarkar and Solanki 2011).

Indian *Alstonia scholaris*, sometimes called as Sitwan Chaal or milky pine acts as an expectorant and anti-tussive within herbal preparations for asthma treatment (Shang et al. 2010).

An active compound called Andrographolide within *Andrographis paniculata* (Indian Echinacea) reduces several symptoms associated with bronchial asthma by suppressing the NF-kappa B activity (Bao et al. 2009).

In Brazil, a medicinal plant named *Amburana cearensis* has traditionally been used to manage majority of the respiratory tract illnesses (Carvalho et al. 2012).

In India, the medicinal herb *Astercantha longifolia* is the source of Kulikhara, an Ayurvedic drug which serves as an expectorant and anti-tussive agent against bronchial asthma (Singh and Handa 1995).

Allium cepa (Onion), popularly called as Pyaaz in India along with its culinary uses, has also been used as an expectorant and anti-tussive agent in several forms (onion paste and onion cloves) to treat bronchial asthma (Kumar et al. 2010a, b).

The root extracts of Indian *Acacia catechu* (Black catechu or Khair) serve as an expectorant and anti-tussive agents in many herbal preparations against bronchial asthma (Patel and Patel 2019).

Herbal preparations made from tree bark of Indian Albizia lebbeck (Pit Shirish shirisha or Albizia) have been found superiorly effective for treating bronchial asthma (Kumar et al. 2010a, b).

Spasmodic asthma has been found to be effectively treated by *Atropa belladonna*, popularly known as Devils Cherries or simply Belladonna (Khan 2015).

Indian *Acalypha indica* (Kuppi) effectively treats asthma and chronic bronchitis (Zahidin et al. 2017).

In India, the roots and stems of *Bacopa monnieri* L (*Scrophulariaceae*) provide expectorant and anti-tussive properties to the herbal formulations for treating bronchial and chronic asthma (Mohapatra and Rath 2005).

The Indian Olibanum tree, scientifically termed as *Boswellia serrata* possess anti-inflammatory tri-terpenoidic compounds named as boswellic acids. These acids have anti-histaminic properties that severely reduce symptoms of bronchial and chronic asthma (Gupta et al. 1998).

The stem, roots, and bark extracts of Indian *Benincasa hispida* (Ash gourd) serve as an expectorant and anti-tussive agent in many herbal preparations that manage respiratory congestion disorders (Gupta and Premavalli 2010).

Root extracts of *Blumea lacera*, popularly called Kukurmutta or Janglimuli has been traditionally used to manage severe attacks of acute asthma (Verma and Shrivastava 2015). The bronchodilation property imparted by *Cuminum cyminum* (Jeera) clears the phlegm blocks in respiratory tract, thereby substantially reducing the symptoms of asthma (Singh et al. 2017).

Dried extracts of leaf, root, and bark of *Cinnamomum cassia*, popularly termed as Dalchini or Chinese cinnamon promotes clearance of phlegm and unnecessary fluids from bronchial tracts. This helps in effective management of asthma (Savithramma et al. 2007).

Medicinal herbs like *Casuarina equisetifolia*, which are commonly prevalent around the entire coastline of India, have been predominantly used to treat asthma and other respiratory illnesses (Aher et al. 2009).

In India, the commonly termed Bharangi, scientifically called as *Clerodendrum Serratum*, has traditionally been used to treat asthmatic patients by virtue of its treatment of histaminic, inflammatory, and chest decongestive active compounds (Savithramma et al. 2007).

Cissus quadrangularis (popularly called Hajora or Asthisanhari) and *Curcuma longa* (common name: Turmeric or Haldi) have anti-inflammatory and anti-asthmatic characteristics, thereby both play critical roles in asthma treatment (Gupta and Verma 1990).

Callicarpa Macrophylla, also called Beauty Berry is useful in rheumatoid arthritis and asthma (Prasad et al. 2009).

Datura metel, also called Thorn Apple possesses anti-asthmatic, antispasmodic, anti-tussive, and bronchodilator activities (Al-Snafi 2017).

The traditional Chinese medicinal herb called *Ephedra sinica* (popularly termed in China as Ma Huang) contains ephedra alkaloids (ephedrine and pseudoephedrine) that induce the α-adrenergic receptors and β-adrenergic receptors to over secrete nor-epinephrine (adrenaline). Such over-stimulation of adrenaline dilates the respiratory tract and provides symptomatic relief to asthmatic patients (Mei et al. 2016).

In India, Mulethi, scientifically called as *Glycyrrhiza glabra* is used customarily in preparations of herbal formulations for asthma treatment (Ram et al. 2006).

The most common spasmodic respiratory conditions like bronchial inflammation and asthma have been effectively treated by *Grindelia camporum* (commonly called the Great valley gum plant) (Smiroldo 2005).

Lobelia inflate (popularly termed as Indian Tobacco) has been reported to possess active compounds to manage spasmodic asthma and bronchial inflammation (Stansbury et al. 2013).

Myrica esculenta (Kaiphal) is used in treatment of bronchitis and asthma in ayurvedic medicine system (Patel et al. 2010).

Tamarindus indica, commonly known as Tamarind possesses anti-inflammatory compounds that quickly soothe the inflamed respiratory tract, and thereby help in acute asthma treatment (Tayade et al. 2009).

Terminalia belerica, called as Vibhitaki in Ayurveda is composed of several active compounds that reveal anti-spasmodic, anti-asthmatic, and anti-tussive characteristics when used in the preparation of herbal formulations for treating asthma (Kale et al. 2010).

Tragia involucrata, popularly called Bichuti, clears the phlegm and cough of trachea to provide instant relief for asthma patient (Yadav et al. 2015).

Apart from asthma, the anti-tussive and chest decongesting characteristics of *Tussilago farfara* help to treat several respiratory challenges like smoker's cough and emphysema (Wu et al. 2016).

Tinospora cordifolia (also known as Guduchi) has been traditionally known to manage spasms associated with bronchial asthma (Guo et al. 2007).

Indian *Ulmus rubra* (popularly termed as the slippery elm) acts as an expectorant and anti-tussive agent who effectively decongest clogged respiratory tract and relieves most of the asthmatic patients (Prasad et al. 2009).

Withania somnifera, fondly called the Indian Ginseng possesses the ability to substantially inhibit bronchial inflammation and provide clear passage to stagnant bronchial fluids in trachea (Singh et al. 2010; Rayees et al. 2014).

Xanthoxylum nepalensis, most commonly called the Prickly Ash or the Toothache Tree has been successful in managing the symptoms of chronic bronchial asthma (Vashisth et al. 2017).

Zingiber officinale (Ginger) is a powerful natural expectorant to treat coughs, colds, and chronic bronchitis (Khan et al. 2015).

Along with the above most commonly used traditional medicinal plants for treating asthma, the root, stem, plant, bark, fruit extracts of some other medicinal plants like *Sida cordifolia* (Flannel weed) (Franzotti et al. 2000), *Solanum melongena* (Eggplant/Brinjal) (Das and Barua 2013), *Saussurea lappa* (Kut Root) (Pandey et al. 2007), *Sphaeranthus indicus* (Sarpate et al. 2009). *Piper longum* (Pippali) (Dahanukar et al. 1984) *Pimpinella anisum* (Anise) (Javadi et al. 2017) *Phymatodes scolopendria* (Ramanitrahasimbola et al. 2005) *Calotropis procera* (Madaar or Dead Sea Apple or Arka) (Meena et al. 2011), *Cynodon dactylon* (Dhub) (Nagori and Solanki 2011) *Cassia sophera* (caesalpiniaceae) (Nagore et al. 2009) used in effectively treat chronic and acute bronchial asthma.

Chapter 13
Asthma Chinese Herbal Remedies

Traditional Chinese Medicine (TCM) has been considered as one of world's most primitive and oldest medical practices. According to TCM, the treatment of Asthma includes the application of medicinal herbs that manage both the acute and chronic asthma. The TCM has several documented strategies that are involved in successful management of asthma, but we have identified and enlisted mainly some common Chinese herbal remedies for asthma management.

13.1 Anti-asthma Herbal Medicine Intervention (ASHMI)

As the name suggests, ASHMI is a herbal intervention to treat asthma. ASHMI is the method of intervention that includes a herbal formulation containing three-herb concoction, which itself is derived from a 14-herb formula. ASHMI has yielded promising results in both in vitro and in vivo setting (Dhawan et al. 2003; Li et al. 2000). Studies have revealed ASHMI to successfully treat asthma in animal models (Li et al. 2000). Randomized clinical trials with the standard-of-care therapy (Oral Prednisone) concluded the ASHMI intervention to be safe and highly effective (Wen et al. 2005). Additional animal studies involving ASHMI reported a wide range of beneficial effects by its direct modulation of airway reactivity and associated inflammation (Srivastava et al. 2005).

13.2 Modified Mai Men Dong Tang (mMMDT)

The Modified Mai Men Dong Tang derived from the original Mai Men Dong Tang (MMDT)—the herbal formulation, specifically and efficiently treated bronchial asthma. In comparison to MMDT, the modified MMDT is a five-herbal (Lantern

S. Rayees, I. Din, *Asthma: Pathophysiology, Herbal and Modern Therapeutic Interventions*, SpringerBriefs in Immunology, https://doi.org/10.1007/978-3-030-70270-0_13

Tridax, Licorice, Ophiopogon, American ginseng, and Pinellia) formulation. The immunological superiority provided by mMMDT formulation in animal models provides substantial evidence for its usage in humans (Hsu et al. 2005). Research undertaken by Hsu et al. (2005) revealed that mMMDT may find usage in chronic and acute asthma modalities.

13.3 Ding Chuan Tang

A popular Chinese herbal formulation named Din Chuan Tang (DCT) has long been used as chest decongestant. Composition of Din Chuan Tang involves concoction of nine Chinese medicinal herbs including *Prunus armeniaca, Perilla frutescens, Scutellaria baricalensis, Morus alba, Glycyrrhiza uralensis, Tussilago farfara, Ephedra sinica, Ginkgo biloba,* and *Pinellia ternate* (Huntley and Ernst 2000).

13.4 STA-1

STA-1 is a herbal formulation made of Mai-Men-Dong-Tang and Lui-Wei-Di-Huang-Wan. Composition STA-1 includes Chinese herbs like *Radix Ophiopogonis, Tuber Pinellia, Radix Panacis Quinquefolii, Radix Glycyrrhizae, Radix Rehmanniae Preparata, Cortex Moutan Radicis, Fructus Corni, Poria, Rhizoma Alismatis,* and *Radix Dioscoreae* (Chang et al. 2006). This study supported that STA-1 plays a critical role in successful treatment of chronic and acute asthma.

13.5 Sophora flavescens Ait

An oriental medicinal herb named *Sophora flavescens* (Ku Shen) has been useful in asthma treatment, respiratory ailments, and chest congestion disorders. Study undertaken by Hoang et al. (2007) concluded that *Sophora flavescens* displayed acceptable safety profile and effectively reversed the damage induced by fibrosis, without any short-term and long-term adverse effects in asthmatic patients (Hoang et al. 2007).

Other than the above herbal remedies, following herbal remedies are equally effective to be used in the management of asthma:

13.6 Ma Huang

Ephedra sinica or *Ma huang*, is a famous conventional Chinese herb and studied in Chinese literature. Its treatment is useful in asthma, hay fever, hemorrhage, cough, circulatory collapse, nasal congestion, and shock (Chen and Hu 2010).

13.7 Minor Blue Dragon

This is a herbal product combination/formulation used in allergy, cough, and asthma (Bielory et al. 2004).

13.8 *Ginkgo biloba*

This is a worldwide used herbal plant against several health inconveniences such as coronary angina, hypercholesterolemia, artery disease, and Parkinson's treatment. In Europe, it is generally used for its anti-asthmatic effects and circulatory disorders. Chinese traditional medicine mentions that it expels phlegm and reduced wheezing, other antibiotic and anti-tubercular advantages (Babayigit et al. 2009).

13.9 Ginseng

Panax ginseng is important herb known as "dose of immortality" used in a variety of health disorders. It is explained by herbal physicians as dominant "adaptogen" which assists the body to cope with stress. Ginseng is considered as a herb, which tones lungs and augments *qi* (energy) for labored breathing with exertion, wheezing, shortness/breath as per the Chinese traditional medicine. Its leaves are used as emetics, expectorants, and root used as cough and an antiseptic. The herb comprises of antioxidants, volatile oil, peptides, polysaccharides, saponins, alcohols, fatty acids, and vitamins (Babayigit et al. 2008).

13.10 Liquorice

It is commonly known as *Glycyrrhiza glabra*. Its root is used as cough expectorant for bronchitis and asthma. It has been demonstrated as a herb that stops coughing, moistens lungs, and wheezing as per the Chinese traditional medicine. It tonifies *qi* in apricot combination, another helpful effect attributed to antibacterial liquorice (Houssen et al. 2010).

13.11   *Zizyphi fructus* and *Zizyphus jujuba*

Zizyphi fructus and *Zizyphus jujuba* var. *Spinosa* are main components of several herbal Chinese anti-asthmatic treatments. They are also considered to have hypnotic and tranquilizing effects. They are also used for several other ailments such as diarrhea, dysentery, colic, dysuria, rheumatism, hepatitis, gingivitis, tuberculosis, and gonorrhea, (Lee et al. 2012).

13.12   Cinnamon

This plant is used for allergies, asthma, and several other health problems and has been documented to have counterirritant effects and used in rheumatic situation as well. Small doses stimulate respiration; treat lung congestion, asthma, bronchitis, and rhinitis. Chinese traditional medicine calls cinnamon an acrid herb and used for edema outcomes from accumulation of phlegm (Gunawardena et al. 2014).

13.13   Blipleuru

Bupleurum falcatum roots are used in Chinese herbal-based treatment for asthma and allergies. *Bupleurum chinense* has anti-tussive, analgesic, sedative, antipyretic, and anti-inflammatory activity showing mild side effects of infrequent nausea and vomiting. *Bupleurum* comprises of fatty acids, glycosides, and saikosaponins (bupleurumol, adonitol, spinasterol, arginine, acids like oleic, palmitic, linolenic, lignoceric, stearic, saikosaponins a, c, and d, daikogenin, isorhamnetin, rutin, longispinogenin, isoquercetin, quercetin, and narcissi) (Sim et al. 2017). Possibly, most of its pharmacological activities may be due to synergistic effects of these components.

13.14   Sckisandra Ckintvisis

Sckisandra Ckintvisis dried fruit are used in various anti-asthmatic herbal preparations which contain sesquicarene, β-chamigrene, β-2-bisabolene, α-ylangene, and glycosides. *Schisandra* stimulates respiratory center and has anti-tussive and expectorant properties (Chen et al. 2014).

13.15   *Ligusticum wallichii*

It is anti-asthmatic herb known as Senkyu in Japan and Chuan Xiong in China. Chinese medicine documented that it invigorates blood and has antibiotic properties. It is composed of neocnidilide, ferulic acid, cnidilide, and ligustilide (Shao et al. 1994).

13.16   Other Herbs

The Chinese herbal medicines are prepared by several herbal combinations.

Apricot dried kernel *(Prunus armeniaca)* is useful in several combinations of herbal contents amygdalase and glucoside amygdalin. Pepsin and amygdalasein of stomach hydrolyze amygdalin to prepare cyanic acid, which stimulates respiration center, producing anti-asthmatic and anti-tussive properties (Hwang et al. 2008).

Pinellia ternata (hange) antiemetic drug has anti-tussive and expectorant properties, used in dampness and resolve phlegm removal in traditional medicine, contains essential oils, choline, amino acids, and L-ephedrine (Lee et al. 2013).

Asiari (Corbicula japonica) is explained as herb in traditional medicine which induces diaphoresis, relieves pain, dispels cold, warms the lungs, and does phlegm reduction (Kim et al. 2004).

Trichosanthes kirilowii has expectorant, anti-tussive, and laxative properties and has been declared to alter phlegm and has antibiotic properties (Dat et al. 2010) possibly due to components such as resin, organic acids, and saponin.

Perilla frutescens (soslii) has several pharmacological effects like diaphoretic, expectorant, anti-tussive, sedative, and antispasmodic. It is also used for cephalic pulmonary and uterine conditions. It is traditional medicine for cephalgia, asthma, cholera, bronchitis, chest, fish poison, cough, malaria, colds, flu, nausea, rheumatism, pregnancy, sunstroke, spasm, and uteritis (Makino et al. 2001).

Dioscorea nipponica is documented as a herb in traditional Chinese herbal medicine for eliminating cold, expelling phlegm, pain relieving, and reducing cough (Junchao et al. 2017).

We further compiled a list of several important studies where the researchers characterized the active constituent(s) responsible for the anti-asthma activity already known herbal plant (Table 13.1):

Furthermore, given below (Table 13.2) is the compiled list of several representative studies that demonstrated the anti-asthma potential of new herbal plants:

Table 13.1 List of plants known for anti-asthma properties or management of Asthma

SN	Name of plant	Part used/extract/fraction	Major chemical constituent	References
1	*Adhatoda vasica* Nees	Leaves, roots	Alkaloids	Paliwa et al. (2000)
2	*Elaeocarpus sphaericus*	Fruits/aqueous, pet-ether, benzene, acetone, and ethanol	Glycoside, steroids, alkaloid, flavanoids	Singh et al. (2000)
3	*Galphimia glauca*	Aerial/alcohol extract/ ethyl-acetate	Tetragalloylquinic acid, quercetin	Campos et al. (2001)
4	*Mikania glomerata*	Leaves/aqueous, hydroalcohol	Coumarin	de Moura et al. (2002)
5	*Lepidium sativum*	Seeds/ethanol fractions	Alkaloids, flavonoids	Mali et al. (2008)
6	*Passiflora incarnata*	Fruits/methanol	alkaloids	Dhawan et al. (2003)
7	*Pavetta crassipes*	Leaves/aqueous	Flavonoids, tannins, anthraquinones	Amos et al. (1998)
8	*Tephrosia purpurea*	Aerial parts/ethanol extract	Flavonoids, tephrosin	Gokhale et al. (2000)
9	*Gleditsia sinensis* Lam.	Leaves/decoctions	Carrageenin and croton oil	Dai et al. (2002)

Table 13.2 List of recent studies which demonstrated promising anti-asthma effect of some herbs plants

Sr. no	Authors	Title of the study	Year of publication
1	MS Youssouf et al.	Anti-anaphylactic effect of *Euphorbia hirta*	2007
2	Ming-Chun Wen et al.	Efficacy and tolerability of anti-asthma herbal medicine intervention in adult patients with moderate-severe allergic asthma	2005
3	P.A. Akah et al.	Evaluation of the anti-asthmatic property of *Asystasia gangetica* leaf extracts	2003
5	M.H. Boskabady	Anti-asthmatic effect of *Nigella sativa* in airways of asthmatic patients	2010
6	Ju-Young Jung et al.	Antioxidant and anti-asthmatic effects of saucerneol D in a mouse model of airway inflammation	2011
7	Singh SK et al.	A review on anti-asthmatic activity of traditional medicinal plants	2014
8	Matsuda H et al.	Anti-allergic effects of Cnidii Monnieri Fructus (Dried Fruits of *Cnidium monnieri*) and its major component, Osthol	2002
9	Aher A N et al.	Evaluation of anti-histaminic activity of *Casuarina equisetifolia* frost (casuarinaceae)	2009
10	Ae-Ran Kim et al.	Acupuncture treatment of a patient with persistent allergic rhinitis complicated by rhinosinusitis and asthma	2011
11	Bhalke R.D and Gosavi S.A	Anti-stress and anti-allergic effect of *Argemone Mexicana* Stems in Asthma	2009
12	Shakarami Z et al.	Evaluation of the protective and therapeutic effects of *Pistacia atlantica* gum aqueous extract on cellular and pathological aspects of experimental asthma in Balb/c mice	2019
13	Jayaprakasam B et al.	Constituents of the anti-asthma herbal formula ASHMI™ synergistically inhibit IL-4 and IL-5 secretion by murine Th2 memory cells, and eotaxin by human lung fibroblasts in vitro	2013
14	Monteiro TM et al.	Anti-asthmatic and anxiolytic effects of Herissantia tiubae, a Brazilian medicinal plant	2016
15	Zhang T et al.	Pharmacology and immunological mechanisms of an herbal medicine, ASHMI™ on allergic asthma	2010

Chapter 14
Future Potential of Herbal-Based Medicinal Treatment for Management of Asthma

Plant-based medicine, sometimes called as phytomedicine has been successful in effective treatment of asthma (Aher et al. 2009). The market for phytomedicine has always been huge, where close to three-fourth of the prescribed and over-the-counter drugs are prepared from natural and plant-based resources (Shakarami et al. 2019). Respiratory ailments have long been an area of active research and several plant-based formulations have provided complete, successful, long-term symptomatic relief. In summary, the future of phytomedicine towards the treatment of asthma holds immense potential.

© The Author(s), under exclusive licence to Springer Nature Switzerland AG 2021   51
S. Rayees, I. Din, *Asthma: Pathophysiology, Herbal and Modern Therapeutic Interventions*, SpringerBriefs in Immunology,
https://doi.org/10.1007/978-3-030-70270-0_14

References

Achola KJ, Munenge RW (1998) Bronchodilating and uterine activities of Ageratum conyzoides extract. Pharm Biol 36(2):93–96

Afkarian M, Sedy JR, Yang J, Jacobson NG, Cereb N, Yang SY, Murphy TL, Murphy KM (2002) T-bet is a STAT1-induced regulator of IL-12R expression in naive CD4+ T cells. Nat Immunol 3:549–557

Agache I, Ciobanu C, Agache C, Anghel M (2010) Increased serum IL-17 is an independent risk factor for severe asthma. Respir Med 104(8):1131–1137

Agertoft L, Pedersen S (2000) Effect of long-term treatment with inhaled budesonide on adult height in children with asthma. N Engl J Med 343(15):1064–1069

Agrawal A, Gosh B (2011) Why is asthma on the increase: nurture via nature. Sci Cult 2011

Aher AN, Pal SC, Patil UK, Yadav SK, Bhattacharya S (2009) Evaluation of anthistaminic activity of Casuarina equisetifolia frost (Casuarinaceae). Pharmacologyonline 1:1144–1149

Akah PA, Ezike AC, Nwafor SV, Okoli CO, Enwerem NM (2003) Evaluation of the anti-asthmatic property of Asystasia gangetica leaf extracts. J Ethnopharmacol 89(1):25–36

Al-Snafi AE (2017) Medical importance of Datura fastuosa (syn: Datura metel) and Datura stramonium-a review. IOSR J Pharm 7(2):43–58

Ammon HPT (2006) Boswellic acids in chronic inflammatory diseases. Planta Med 72(12):1100–1116

Amos S, Gamaniel K, Akah P, Wambebe C, Okwuasaba FK (1998) Anti-inflammatory and muscle relaxant effects of the aqueous extract of Pavetta crassipes leaves. Fitoterapia (Milano) 69(5):425–429

Anderson GP (2008) Endotyping asthma: new insights into key pathogenic mechanisms in a complex, heterogeneous disease. Lancet 372:1107–1119

Andrew Mcivor R, Pizzichini E, Turner MO, Hussack P, Hargreave FE, Sears MR (1998) Potential masking effects of salmeterol on airway inflammation in asthma. Am J Respir Crit Care Med 158(3):924–930

Anil K, Ramu P (2002) Effect of methanolic extract of Benincasa hispida against histamine and acetylcholine induced bronchospasm in Guinea pigs. Indian J Pharmacol 34(5):365

Ashkar S, Weber GF, Panoutsakopoulou V, Sanchirico ME, Jansson M, Zawaideh S, Rittling SR, Denhardt DT, Glimcher MJ, Cantor H (2000) Eta-1 (osteopontin): an early component of type-1 (cell-mediated) immunity. Science 287:860–864

Ayres JG, Higgins B, Chilvers ER, Ayre G, Blogg M, Fox H (2004) Efficacy and tolerability of anti-immunoglobulin E therapy with omalizumab in patients with poorly controlled (moderate-to-severe) allergic asthma. Allergy 59:701–708

© The Author(s), under exclusive licence to Springer Nature Switzerland AG 2021   53
S. Rayees, I. Din, *Asthma: Pathophysiology, Herbal and Modern Therapeutic Interventions*, SpringerBriefs in Immunology,
https://doi.org/10.1007/978-3-030-70270-0

Babayigit A, Olmez D, Karaman O, Bagriyanik HA, Yilmaz O, Kivcak B, Erbil G, Uzuner N (2008, September) Ginseng ameliorates chronic histopathologic changes in a murine model of asthma. Allergy Asthma Proc 29(5):493

Babayigit A, Olmez D, Karaman O, Ozogul C, Yilmaz O, Kivcak B, Erbil G, Uzuner N (2009) Effects of Ginkgo biloba on airway histology in a mouse model of chronic asthma. Allergy Asthma Proc 30(2):186–191

Bagnasco D, Ferrando M, Varricchi G, Passalacqua G, Canonica GW (2016) A critical evaluation of anti-IL-13 and anti-IL-4 strategies in severe asthma. Int Arch Allergy Immunol 170(2):122–131

Balzano G, Fuschillo S, Gaudiosi C (2002) Leukotriene receptor antagonists in the treatment of asthma: an update. Allergy 57:16–19

Bao Z, Guan S, Cheng C, Wu S, Wong SH, Kemeny DM, Leung BP, Wong WF (2009) A novel antiinflammatory role for andrographolide in asthma via inhibition of the nuclear factor-κB pathway. Am J Respir Crit Care Med 179(8):657–665

Barnes PJ (2003) Theophylline: new perspectives for an old drug. Am J Resp Crit Care Med 167(6):813–818

Barnes PJ (2013) Theophylline. Am J Respir Crit Care Med 188:901–906

Barnes PJ (1996) Pathophysiology of asthma. Br J Clin Pharmacol 42:3–10

Barnes PJ (1998a) Efficacy of inhaled corticosteroids in asthma. J Allergy Clin Immunol 102(4):531–538

Barnes PJ (1998b) Pharmacology of airway smooth muscle. Am J Respir Crit Care Med 158:S123–S132

Barnes PJ (2006) Transcription factors in airway diseases. Lab Investig 86:867–872

Barnes PJ (2010) New therapies for asthma: is there any progress? Trends Pharmacol Sci 31:335–343

Barnes PJ, Adcock IM (2003) How do corticosteroids work in asthma? Ann Intern Med 139(1):359–370

Barnes PJ, Pauwels RA (1994) Theophylline in the management of asthma: time for reappraisal? Eur Respir J 7(3):579–591

Barr RG, Rowe BH, Camargo CA (2003) Methylxanthines for exacerbations of chronic obstructive pulmonary disease: meta-analysis of randomised trials. BMJ 327:643

Barrett NA, Austen KF (2009) Innate cells and T helper 2 cell immunity in airway inflammation. Immunity 31:425–437

Beghe B, Rabe KF, Fabbri LM (2013) Phosphodiesterase-4 inhibitor therapy for lung diseases. Am J Respir Crit Care Med 188:271–278

Bel EH (2013) Clinical practice. Mild asthma. N Engl J Med 369:549–557

Bettelli E, Carrier Y, GAO W, Korn T, Strom TB, Oukka M, Weiner HL, Kuchroo VK (2006) Reciprocal developmental pathways for the generation of pathogenic effector TH17 and regulatory T cells. Nature 441:235–238

Beuther DA (2010) Recent insight into obesity and asthma. Curr Opin Pulm Med 16:64–70

Bhalke RD, Gosavi SA (2009) Anti-stress and antiallergic effect of Argemone mexicana stems in asthma. Arch Pharm Sci Res 1(1):127–129

Bielory L, Lupoli K (1999) Herbal interventions in asthma and allergy. J Asthma 36(1):1–65

Bielory L, Russin J, Zuckerman GB (2004) Clinical efficacy, mechanisms of action, and adverse effects of complementary and alternative medicine therapies for asthma. Allergy Asthma Proc 25(5):283–291

Bjermer L, Diamant Z (2002) The use of leukotriene receptor antagonists (LTRAs) as complementary therapy in asthma. Monaldi Arch Chest Dis 57(1):76–83

Blanchard C, Mishra A, Saito-Akei H, Monk P, Anderson I, Rothenberg ME (2005) Inhibition of human interleukin-13-induced respiratory and oesophageal inflammation by anti-human-interleukin-13 antibody (CAT-354). Clin Exp Allergy 35:1096–1103

Bleecker E (1998) Inhaled corticosteroids: current products and their role in patient care. J Allergy Clin Immunol 101(2):S400–S402

Blyth DI, Pedrick MS, Savage TJ, Hessel EM, Fattah D (1996) Lung inflammation and epithelial changes in a murine model of atopic asthma. Am J Respir Cell Mol Biol 14:425–438

Boskabady MH, Mohsenpoor N, Takaloo L (2010) Antiasthmatic effect of Nigella sativa in airways of asthmatic patients. Phytomedicine 17(10):707–713

Bousquet J, Aubier M, Sastre J, Izquierdo JL, Adler LM, Hofbauer P, Rost KD, Harnest U, Kroemer B, Albrecht A, Bredenbroker D (2006) Comparison of roflumilast, an oral anti-inflammatory, with beclomethasone dipropionate in the treatment of persistent asthma. Allergy 61:72–78

Bousquet J, Cabrera P, Berkman N, Buhl R, Holgate S, Wenzel S, Fox H, Hedgecock S, Blogg M, Cioppa GD (2005) The effect of treatment with omalizumab, an anti-IgE antibody, on asthma exacerbations and emergency medical visits in patients with severe persistent asthma. Allergy 60:302–308

Bousquet J, Jeffery PK, Busse WW, Johnson M, Vignola AM (2000) Asthma. From bronchoconstriction to airways inflammation and remodeling. Am J Respir Crit Care Med 161:1720–1745

Boyce JA, Austen KF (2005) No audible wheezing: nuggets and conundrums from mouse asthma models. J Exp Med 201:1869–1873

Braman SS (2006) The global burden of asthma. Chest 130:4S–12S

Buhl R (2006) Local oropharyngeal side effects of inhaled corticosteroids in patients with asthma. Allergy 61(5):518–526

Bullens DM, Truyen E, Coteur L, Dilissen E, Hellings PW, Dupont LJ, Ceuppens JL (2006) IL-17 mRNA in sputum of asthmatic patients: linking T cell driven inflammation and granulocytic influx? Respir Res 7(1):135

Busse WW, Lemanske RF, JR. (2001) Asthma. N Engl J Med 344:350–362

Busse WW, Holgate S, Kerwin E, Chon Y, Feng J, Lin J, Lin SL (2013) Randomized, double-blind, placebo-controlled study of brodalumab, a human anti–IL-17 receptor monoclonal antibody, in moderate to severe asthma. Am J Respir Crit Care Med 188(11):1294–1302

Campbell LM (1999) Once-daily inhaled corticosteroids in mild to moderate asthma: improving acceptance of treatment. Drugs 58(Suppl 4):25–33

Campos MG, Toxqui E, Tortoriello J, Oropeza MV, Ponce H, Vargas MH, Montaño LM (2001) Galphimia glauca organic fraction antagonizes LTD4-induced contraction in Guinea pig airways. J Ethnopharmacol 74(1):7–15

Caramori G, Groneberg D, Ito K, Casolari P, Adcock IM, PAPI A (2008) New drugs targeting Th2 lymphocytes in asthma. J Occup Med Toxicol 3(Suppl 1):S6

Caramori G, Ito K, Adcock IM (2004) Transcription factors in asthma and COPD. IDrugs 7:764–770

Carvalho EM, Cunha GHD, Fechine FV, Uchôa CRA, Moraes Filho MOD, Bezerra FAF, Moraes MEAD (2012) Efficacy and safety of Cumaru syrup as complementary therapy in mild persistent asthma: a double-blind, randomized, placebo-controlled study. Braz J Pharm Sci 48(4):629–637

Castillo EF, Zheng H, Yang XO (2018) Orchestration of epithelial-derived cytokines and innate immune cells in allergic airway inflammation. Cytokine Growth Factor Rev 39:19–25

Cazzola M, Donner CF (2000) Long-acting β 2 agonists in the management of stable chronic obstructive pulmonary disease. Drugs 60(2):307–320

Chan CK, Kuo ML, Shen JJ, See LC, Chang HH, Huang JL (2006) Ding Chuan Tang, a Chinese herb decoction, could improve airway hyper-responsiveness in stabilized asthmatic children: a randomized, double-blind clinical trial. Pediatr Allergy Immunol 17(5):316–322

Chang TT, Huang CC, Hsu CH (2006) Clinical evaluation of the Chinese herbal medicine formula STA-1 in the treatment of allergic asthma. Phytother Res 20(5):342–347

Channa S, Dar A, Yaqoob M, Anjum S, Sultani Z (2003) Broncho-vasodilatory activity of fractions and pure constituents isolated from Bacopa monniera. J Ethnopharmacol 86(1):27–35

Chen W, Jin W, Hardegen N, Lei KJ, Li L, Marinos N, Mcgrady G, Wahl SM (2003) Conversion of peripheral CD4+CD25- naive T cells to CD4+CD25+ regulatory T cells by TGF-beta induction of transcription factor Foxp3. J Exp Med 198:1875–1886

Chen X, Huang Y, Feng J, Jiang XF, Xiao WF, Chen XX (2014) Antioxidant and anti-inflammatory effects of Schisandra and Paeonia extracts in the treatment of asthma. Exp Ther Med 8(5):1479–1483

Chen Z, Hu G (2010) Effect of modified shegan mahuang decoction on cytokines in children patients with cough and variant asthma. Zhongguo Zhong xi yi jie he za zhi Zhongguo Zhongxiyi jiehe zazhi= Chin J Integr Trad West Med 30(2):208–210

Cher DJ, Mosmann TR (1987) Two types of murine helper T cell clone. II. Delayed-type hypersensitivity is mediated by TH1 clones. J Immunol 138:3688–3694

Chesné J, Braza F, Mahay G, Brouard S, Aronica M, Magnan A (2014) IL-17 in severe asthma. Where do we stand? Am J Respir Crit Care Med 190(10):1094–1101

Chetta A, Foresi A, Del Donno M, Bertorelli G, Pesci A, Olivieri D (1997) Airways remodeling is a distinctive feature of asthma and is related to severity of disease. Chest 111:852–857

Chiba Y, Todoroki M, Nishida Y, Tanabe M, Misawa M (2009) A novel STAT6 inhibitor AS1517499 ameliorates antigen-induced bronchial hypercontractility in mice. Am J Respir Cell Mol Biol 41:516–524

Childhood Asthma Management Program Research Group (2000) Long-term effects of budesonide or nedocromil in children with asthma. N Engl J Med 343(15):1054–1063

Choby GW, Lee S (2015) Pharmacotherapy for the treatment of asthma: current treatment options and future directions. Int Forum Allergy Rhinol 5(Suppl 1):S35–S40

Christodoulopoulos P, Cameron L, Nakamura Y, Lemiere C, Muro S, Dugas M, Boulet LP, Laviolette M, Olivenstein R, Hamid Q (2001) TH2 cytokine-associated transcription factors in atopic and nonatopic asthma: evidence for differential signal transducer and activator of transcription 6 expression. J Allergy Clin Immunol 107:586–591

Cochrane MG, Bala MV, Downs KE, Mauskopf J, Ben-Joseph RH (2000) Inhaled corticosteroids for asthma therapy: patient compliance, devices, and inhalation technique. Chest 117(2):542–550

Cohn L, Ray A (2000) T-helper type 2 cell-directed therapy for asthma. Pharmacol Ther 88:187–196

Cook ML, Bochner BS (2010) Update on biological therapeutics for asthma. World Allergy Organ J 3:188–194

Corren J (2019) New targeted therapies for uncontrolled asthma. J Allergy Clin Immunol Pract 7(5):1394–1403

Corrigan CJ, Hartnell A, Kay AB (1988) T lymphocyte activation in acute severe asthma. Lancet 1:1129–1132

Cosmi L, Liotta F, Maggi E, Romagnani S, Annunziato F (2011) Th17 cells: new players in asthma pathogenesis. Allergy 66:989–998

Cowburn AS, Sladek K, Soja J, Adamek L, Nizankowska E, Szczeklik A, Lam BK, Penrose JF, Austen FK, Holgate ST, Sampson AP (1998) Overexpression of leukotriene C4 synthase in bronchial biopsies from patients with aspirin-intolerant asthma. J Clin Invest 101:834–846

Creticos PS, Peters SP, Adkinson NF Jr, Naclerio RM, Hayes EC, Norman PS, Lichtenstein LM (1984) Peptide leukotriene release after antigen challenge in patients sensitive to ragweed. N Engl J Med 310:1626–1630

Dahanukar SA, Karandikar SM, Desai M (1984) Effect of *Piper longum* in childhood asthma. Indian Drugs 21(9):384–388

Dahl R (2006) Systemic side effects of inhaled corticosteroids in patients with asthma. Respir Med 100(8):1307–1317

Dai Y, Chan YP, Chu LM, But PPH (2002) Antiallergic and anti-inflammatory properties of the ethanolic extract from Gleditsia sinensis. Biol Pharm Bull 25(9):1179–1182

Das M, Barua N (2013) Pharmacological activities of Solanum melongena Linn.(Brinjal plant). Int J Green Pharm 7(4):274

Dat NT, Jin X, Hong YS, Lee JJ (2010) An isoaurone and other constituents from Trichosanthes kirilowii seeds inhibit hypoxia-inducible factor-1 and nuclear factor-κB. J Nat Prod 73(6):1167–1169

De Boever EH, Ashman C, Cahn AP, Locantore NW, Overend P, Pouliquen IJ, Serone AP, Wright TJ, Jenkins MM, Panesar IS, Thiagarajah SS, Wenzel SE (2014) Efficacy and safety of an anti-IL-13 mAb in patients with severe asthma: a randomized trial. J Allergy Clin Immunol 133(4):989–996

de Moura RS, Costa SS, Jansen JM, Silva CA, Lopes CS, Bernardo-Filho M, da Silva VN, Criddle DN, Portela BN, Rubenich LMS, Araújo RG (2002) Bronchodilator activity of Mikania glomerata Sprengel on human bronchi and Guinea-pig trachea. J Pharm Pharmacol 54(2):249–256

Desmet C, Gosset P, Henry E, Garze V, Faisca P, Vos N, Jaspar F, Melotte D, Lambrecht B, Desmecht D, Pajak B, Moser M, Lekeux P, Bureau F (2005) Treatment of experimental asthma by decoy-mediated local inhibition of activator protein-1. Am J Respir Crit Care Med 172:671–678

Dhand R, Dolovich M, Chipps B, Myers TR, Restrepo R, Rosen Farrar J (2012) The role of nebulized therapy in the management of COPD: evidence and recommendations. COPD: J Chron Obstruct Pulmon Dis 9(1):58–72

Dhawan K, Kumar S, Sharma A (2003) Antiasthmatic activity of the methanol extract of leaves of Passiflora incarnata. Phytother Res 17(7):821–822

Drazen JM, Israel E, O'Byrne PM (1999) Treatment of asthma with drugs modifying the leukotriene pathway. N Engl J Med 340(3):197–206

Einarsson O, Geba GP, Zhou Z, Landry ML, Panettieri RA Jr, Tristram D, Welliver R, Metinko A, Elias JA (1995) Interleukin-11 in respiratory inflammation. Ann N Y Acad Sci 762:89–100. discussion 100-1

Ernst P, Spitzer WO, Suissa S, Cockcroft D, Habbick B, Horwitz RI, Boivin JF, McNutt M, Buist AS (1992) Risk of fatal and near-fatal asthma in relation to inhaled corticosteroid use. JAMA 268(24):3462–3464

Erpenbeck VJ, Hagenberg A, Krentel H, Discher M, Braun A, Hohlfeld JM, Krug N (2006) Regulation of GATA-3, c-maf and T-bet mRNA expression in bronchoalveolar lavage cells and bronchial biopsies after segmental allergen challenge. Int Arch Allergy Immunol 139:306–316

Erpenbeck VJ, Hohlfeld JM, Discher M, Krentel H, Hagenberg A, Braun A, Krug N (2003) Increased messenger RNA expression of c-maf and GATA-3 after segmental allergen challenge in allergic asthmatics. Chest 123:370S–371S

Fajt ML, Wenzel SE (2014) Biologic therapy in asthma: entering the new age of personalized medicine. J Asthma 51:669–676

Fallon PG, Emson CL, Smith P, Mckenzie AN (2001) IL-13 overexpression predisposes to anaphylaxis following antigen sensitization. J Immunol 166:2712–2716

Fan Chung K (2006) Phosphodiesterase inhibitors in airways disease. Eur J Pharmacol 533:110–117

Farahani R, Sherkat R, Hakemi MG, Eskandari N, Yazdani R (2014) Cytokines (interleukin-9, IL-17, IL-22, IL-25 and IL-33) and asthma. Adv Biomed Res 3:127

Finnerty JP, Wood-Baker R, Thomson H, Holgate ST (1992) Role of leukotrienes in exercise-induced asthma. Am Rev Respir Dis 145:746–749

Finotto S, De Sanctis GT, Lehr HA, Herz U, Buerke M, Schipp M, Bartsch B, Atreya R, Schmitt E, Galle PR, Renz H, Neurath MF (2001) Treatment of allergic airway inflammation and hyperresponsiveness by antisense-induced local blockade of GATA-3 expression. J Exp Med 193:1247–1260

Fish JE, Peters SP (1999) Airway remodeling and persistent airway obstruction in asthma. J Allergy Clin Immunol 104:509–516

Franciosi LG, Diamant Z, Banner KH, Zuiker R, Morelli N, Kamerling IM, de Kam ML, Burggraaf J, Cohen AF, Cazzola M, Calzetta L, Singh D, Spina D, Walker MJ, Page CP (2013) Efficacy and safety of RPL554, a dual PDE3 and PDE4 inhibitor, in healthy volunteers and in patients with asthma or chronic obstructive pulmonary disease: findings from four clinical trials. Lancet Respir Med 1:714–727

Franzotti EM, Santos CVF, Rodrigues HMSL, Mourao RHV, Andrade MR, Antoniolli AR (2000) Anti-inflammatory, analgesic activity and acute toxicity of Sida cordifolia L. (Malva-branca). J Ethnopharmacol 72(1–2):273–277

Fuso L, Mores N, Valente S, Malerba M, Montuschi P (2013) Long-acting beta-agonists and their association with inhaled corticosteroids in COPD. Curr Med Chem 20(12):1477–1495

Gallelli L, Busceti MT, Vatrella A, Maselli R, Pelaia G (2013) Update on anticytokine treatment for asthma. Biomed Res Int 2013:104315

Gasser P, Tarchevskaya SS, Guntern P, Brigger D, Ruppli R, Zbären N, Kleinboelting S, Heusser C, Jardetzky ST, Eggel A (2020) The mechanistic and functional profile of the therapeutic anti-IgE antibody ligelizumab differs from omalizumab. Nat Commun 11(1):1–14

Gauvreau GM, O'Byrne PM, Boulet LP, Wang Y, Cockcroft D, Bigler J et al (2014) Effects of an anti-TSLP antibody on allergen-induced asthmatic responses. N Engl J Med 370(22):2102–2110

Gavett SH, Chen X, Finkelman F, Wills-Karp M (1994) Depletion of murine CD4+ T lymphocytes prevents antigen-induced airway hyperreactivity and pulmonary eosinophilia. Am J Respir Cell Mol Biol 10:587–593

Ghaffar O, Christodoulopoulos P, Lamkhioued B, Wright E, Ihaku D, Nakamura Y, Frenkiel S, Hamid Q (2000) In vivo expression of signal transducer and activator of transcription factor 6 (STAT6) in nasal mucosa from atopic allergic rhinitis: effect of topical corticosteroids. Clin Exp Allergy 30:86–93

Gibeon D, Menzies-Gow A (2013) Recent changes in the drug treatment of allergic asthma. Clin Med 13:477–481

Glimcher LH, Murphy KM (2000) Lineage commitment in the immune system: the T helper lymphocyte grows up. Genes Dev 14:1693–1711

Gokhale AB, Dikshit VJ, Damre AS, Kulkarni KR, Saraf MN (2000) Influence of ethanolic extract of Tephrosia purpurea Linn. on mast cells and erythrocytes membrane integrity. Indian J Exp Biol 38(8):837–840

Goswami R, Kaplan MH (2011) A brief history of IL-9. J Immunol 186(6):3283–3288

Govindan S, Viswanathan S, Vijayasekaran V, Alagappan R (1999) A pilot study on the clinical efficacy of Solanum xanthocarpum and Solanum trilobatum in bronchial asthma. J Ethnopharmacol 66(2):205–210

Graham DM, Blaiss MS (2000) Complementary/alternative medicine in the treatment of asthma. Ann Allergy Asthma Immunol 85(6):438–449

Grainge CL, Lau LC, Ward JA, Dulay V, Lahiff G, Wilson S, Holgate S, Davies DE, Howarth PH (2011) Effect of bronchoconstriction on airway remodeling in asthma. N Engl J Med 364:2006–2015

Grandordy BM, Mak JC, Barnes PJ (1994) Modulation of airway smooth muscle beta-adrenoceptor function by a muscarinic agonist. Life Sci 54:185–191

Griffiths B, Ducharme FM (2013) Combined inhaled anticholinergics and short-acting beta 2-agonists for initial treatment of acute asthma in children. Cochrane Database Syst Rev, (8):CD000060

Guhan AR, Cooper S, Oborne J, Lewis S, Bennett J, Tattersfield AE (2000) Systemic effects of formoterol and salmeterol: a dose-response comparison in healthy subjects. Thorax 55(8):650–656

Guilbert TW, Morgan WJ, Zeiger RS, Mauger DT, Boehmer SJ, Szefler SJ et al (2006) Long-term inhaled corticosteroids in preschool children at high risk for asthma. N Engl J Med 354(19):1985–1997

Gunawardena D, Govindaraghavan S, Münch G (2014) Anti-inflammatory properties of cinnamon polyphenols and their monomeric precursors. In: Polyphenols in human health and disease. Academic, Cambridge, pp 409–425

Guo R, Pittler MH, Ernst E (2007) Herbal medicines for the treatment of allergic rhinitis: a systematic review. Ann Allergy Asthma Immunol 99(6):483–495

Gupta A, Prajapati PK (2010) A clinical review of different formulations of vasa (Adhatoda vasica) on Tamaka Shwasa (asthma). Ayu 31:520–524

Gupta I, Gupta V, Parihar A, Gupta S, Lüdtke R, Safayhi H, Ammon HP (1998) Effects of Boswellia serrata gum resin in patients with bronchial asthma: results of a double-blind, placebo-controlled, 6-week clinical study. Eur J Med Res 3(11):511–514

Gupta MM, Verma RK (1990) Unsymmetric tetracyclic triterpenoid from Cissus quadrangularis. Phytochemistry 29(1):336–337

Gupta P, Premavalli KS (2010) Effect of particle size reduction on physicochemical properties of ashgourd (Benincasa hispida) and radish (Raphanus sativus) fibres. Int J Food Sci Nutr 61(1):18–28

Hacha J, Tomlinson K, Maertens L, Paulissen G, Rocks N, Foidart JM, Noel A, Palframan R, Gueders M, Cataldo DD (2012) Nebulized anti-IL-13 monoclonal antibody Fab' fragment reduces allergen-induced asthma. Am J Respir Cell Mol Biol 47:709–717

Hambly N, Nair P (2014) Monoclonal antibodies for the treatment of refractory asthma. Curr Opin Pulm Med 20:87–94

Harris JM, Cabanski CR, Scheerens H, Samineni D, Bradley MS, Cochran C, Staubach P, Metz M, Sussman G, Maurer M (2016) A randomized trial of quilizumab in adults with refractory chronic spontaneous urticaria. J Allergy Clin Immunol 138(6):1730–1732

Hay DW, Torphy TJ, Undem BJ (1995) Cysteinyl leukotrienes in asthma: old mediators up to new tricks. Trends Pharmacol Sci 16:304–309

Henderson WR Jr (1994) The role of leukotrienes in inflammation. Ann Intern Med 121:684–697

Ho IC, Tai TS, Pai SY (2009) GATA3 and the T-cell lineage: essential functions before and after T-helper-2-cell differentiation. Nat Rev Immunol 9:125–135

Hoang BX, Shaw DG, Levine S, Hoang C, Pham P (2007) New approach in asthma treatment using excitatory modulator. Phytother Res 21(6):554–557

Hodsman P, Ashman C, Cahn A, De Boever E, Locantore N, Serone A, Pouliquen I (2013) A phase 1, randomized, placebo-controlled, dose-escalation study of an anti-IL-13 monoclonal antibody in healthy subjects and mild asthmatics. Br J Clin Pharmacol 75(1):118–128

Holgate ST (2014) New strategies with anti-IgE in allergic diseases. World Allergy Organ J 7:17

Holgate ST, Bradding P, Sampson AP, is supported by the Frances, S., & Foundation, A. N (1996) Leukotriene antagonists and synthesis inhibitors: new directions in asthma therapy. J Allergy Clin Immunol 98(1):1–13

Holgate ST, Davies DE, Lackie PM, Wilson SJ, Puddicombe SM, Lordan JL (2000) Epithelial-mesenchymal interactions in the pathogenesis of asthma. J Allergy Clin Immunol 105:193–204

Holmes AM, Solari R, Holgate ST (2011) Animal models of asthma: value, limitations and opportunities for alternative approaches. Drug Discov Today 16:659–670

Hom S, Pisano M (2017) Reslizumab (Cinqair): an Interleukin-5 antagonist for severe asthma of the Eosinophilic phenotype. P T 42(9):564–568

Hooper I (1888) Isoln from Adhatoda vasica Nees, Acanthaceae. J Pharm 18:841–842

Hori S, Nomura T, Sakaguchi S (2003) Control of regulatory T cell development by the transcription factor Foxp3. Science 299:1057–1061

Horie S, Okubo Y, Hossain M, SATO E, Nomura H, Koyama S et al (1997) Interleukin-13 but not interleukin-4 prolong eosinophil survival and induce eosinophil chemotaxis. Intern Med 36(3):179–185

Horiuchi S, Onodera A, Hosokawa H, Watanabe Y, Tanaka T, Sugano S, Suzuki Y, Nakayama T (2011) Genome-wide analysis reveals unique regulation of transcription of Th2-specific genes by GATA3. J Immunol 186:6378–6389

Hoshino A, Tsuji T, Matsuzaki J, Jinushi T, Ashino S, Teramura T, Chamoto K, Tanaka Y, Asakura Y, Sakurai T, Mita Y, Takaoka A, Nakaike S, Takeshima T, Ikeda H, Nishimura T (2004) STAT6-mediated signaling in Th2-dependent allergic asthma: critical role for the development of eosinophilia, airway hyper-responsiveness and mucus hypersecretion, distinct from its role in Th2 differentiation. Int Immunol 16:1497–1505

Houssen ME, Ragab A, Mesbah A, El-Samanoudy AZ, Othman G, Moustafa AF, Badria FA (2010) Natural anti-inflammatory products and leukotriene inhibitors as complementary therapy for bronchial asthma. Clin Biochem 43(10–11):887–890

Hsu CH, Lu CM, Chang TT (2005) Efficacy and safety of modified Mai-men-Dong-Tang for treatment of allergic asthma. Pediatr Allergy Immunol 16(1):76–81

Humbert M, Busse W, Hanania NA, Lowe PJ, Canvin J, Erpenbeck VJ, Holgate S (2014) Omalizumab in asthma: an update on recent developments. J Allergy Clin Immunol Pract 2:525–536.e1

Humbles AA, Reed JL, Parker J, Kiener PA, Molfino NA, Kolbeck R, Coyle AJ (2010) Monoclonal antibody therapy directed against interleukin-9: MEDI-528. In: New drugs and targets for asthma and COPD, vol. 39. Karger Publishers, Basel, pp 137–140

Huntley A, Ernst E (2000) Herbal medicines for asthma: a systematic review. Thorax 55(11):925–929

Hwang HJ, Kim P, Kim CJ, Lee HJ, Shim I, Yin CS, Yang Y, Hahm DH (2008) Antinociceptive effect of amygdalin isolated from Prunus armeniaca on formalin-induced pain in rats. Biol Pharm Bull 31(8):1559–1564

Ichinose M, Barnes PJ (2004) Cytokine-directed therapy in asthma. Curr Drug Targets Inflamm Allergy 3:263–269

Israel E, Cohn J, Dubé L, Drazen JM, Ratner P, Pleskow W et al (1996) Effect of treatment with zileuton, a 5-lipoxygenase inhibitor, in patients with asthma: a randomized controlled trial. JAMA 275(12):931–936

Iwasaki A, Medzhitov R (2004) Toll-like receptor control of the adaptive immune responses. Nat Immunol 5:987–995

Jaffar ZH, Sullivan P, Page C, Costello J (1996) Low-dose theophylline modulates T-lymphocyte activation in allergen-challenged asthmatics. Eur Respir J 9(3):456–462

Jamaleddine G, Diab K, Tabbarah Z, Tawil A, Arayssi T (2002) Leukotriene antagonists and the Churg-Strauss syndrome. Semin Arthritis Theumatism 31(4):218–227. WB Saunders

Jang HY, Ahn KS, Park MJ, Kwon OK, Lee HK, Oh SR (2012) Skullcapflavone II inhibits ovalbumin-induced airway inflammation in a mouse model of asthma. Int Immunopharmacol 12(4):666–674

Javadi B, Sahebkar A, Ahmad Emami S (2017) Medicinal plants for the treatment of asthma: a traditional persian medicine perspective. Curr Pharm Des 23(11):1623–1632

Jayaprakasam B, Yang N, Wen MC, Wang R, Goldfarb J, Sampson H, Li XM (2013) Constituents of the anti–asthma herbal formula ASHMI™ synergistically inhibit IL–4 and IL–5 secretion by murine Th2 memory cells, and eotaxin by human lung fibroblasts in vitro. J Integrat Med 11(3):195

Johnson JR, Wiley RE, Fattouh R, Swirski FK, Gajewska BU, Coyle AJ, Gutierrez-Ramos JC, Ellis R, Inman MD, Jordana M (2004) Continuous exposure to house dust mite elicits chronic airway inflammation and structural remodeling. Am J Respir Crit Care Med 169:378–385

Junchao Y, Zhen W, Yuan W, Liying X, Libin J, Yuanhong Z, Wei Z, Ruilin C, Lu Z (2017) Anti-trachea inflammatory effects of diosgenin from Dioscorea nipponica through interactions with glucocorticoid receptor α. J Int Med Res 45(1):101–113

Jung JY, Lee KY, Lee MY, Jung D, Cho ES, Son HY (2011) Antioxidant and antiasthmatic effects of saucerneol D in a mouse model of airway inflammation. Int Immunopharmacol 11(6):698–705

Kale RN, Patil RN, Patil RY (2010) Asthma and herbal drugs. Pathophysiology 20:4

Kant S (2013) Socio-economic dynamics of asthma. Indian J Med Res 138:446–448

Kaur I, Kumar A, Sharma S (2013) Adhatoda vasica: effect of administering ethanolic extract of adhatoda vasica on blood count and sod enzyme activity against exposure to γ–radiations. Int J Pharm Sci Res 4:4016–4026

Khan AM, Shahzad M, Raza Asim MB, Imran M, Shabbir A (2015) Zingiber officinale ameliorates allergic asthma via suppression of Th2-mediated immune response. Pharm Biol 53(3):359–367

Khan H (2015) Alkaloids: potential therapeutic modality in the management of asthma. J Ayurvedic Herb Med 1(3):3

Kim AR, Choi JY, Kim JI, Jung SY, Choi SM (2011) Acupuncture treatment of a patient with persistent allergic rhinitis complicated by rhinosinusitis and asthma. Evid Based Complement Alternat Med 2011:798081

Kim IG, Kim YI, Hong KE, Yim YK, Lee BR (2004) The effects of Asari Herba Cum Radice (AHCR) herbal acupuncture at St36 on ovalbumin-induced asthma in C57BL mouse. Korean J Acupunct 21(1):61–77

Kim J, Woods A, Becker-Dunn E, Bottomly K (1985) Distinct functional phenotypes of cloned Ia-restricted helper T cells. J Exp Med 162:188–201

Kips JC, Pauwels RA (2001) Long-acting inhaled beta(2)-agonist therapy in asthma. Am J Respir Crit Care Med 164(6):923–932

Kips JC, Pauwels RA (1999) Airway wall remodelling: does it occur and what does it mean? Clin Exp Allergy 29:1457–1466

Kiwamoto T, Ishii Y, Morishima Y, Yoh K, Maeda A, Ishizaki K, Iizuka T, Hegab AE, Matsuno Y, Homma S, Nomura A, Sakamoto T, Takahashi S, Sekizawa K (2006) Transcription factors T-bet and GATA-3 regulate development of airway remodeling. Am J Respir Crit Care Med 174:142–151

Knarborg M, Hilberg O, Dahl R (2014) Methotrexate may be a useful corticosteroid reducing treatment of severe asthma. Ugeskr Laeger 176(15):V10130598

Knight DA, Lim S, Scaffidi AK, Roche N, Chung KF, Stewart GA, THOMPSON PJ (2001) Protease-activated receptors in human airways: upregulation of PAR-2 in respiratory epithelium from patients with asthma. J Allergy Clin Immunol 108:797–803

Kumar D, Prasad DN, Parkash J, Bhatnagar SP, Kumar D (2009) Antiasthmatic activity of ethanolic extract of Aerva lanata Linn. Pharmacologyonline 2:1075–1081

Kumar KS, Bhowmik D, Chiranjib B, Tiwari P (2010a) Allium cepa: A traditional medicinal herb and its health benefits. J Chem Pharm Res 2(1):283–291

Kumar RK, Foster PS (2002) Modeling allergic asthma in mice: pitfalls and opportunities. Am J Respir Cell Mol Biol 27:267–272

Kumar RK, Herbert C, Foster PS (2008) The "classical" ovalbumin challenge model of asthma in mice. Curr Drug Targets 9:485–494

Kumar RK, Herbert C, Webb DC, Li L, Foster PS (2004) Effects of anticytokine therapy in a mouse model of chronic asthma. Am J Respir Crit Care Med 170(10):1043–1048

Kumar S, Bansal P, Gupta V, Sannd R, Rao M (2010b) The clinical effect of Albizia lebbeck stem bark decoction on bronchial asthma. Int J Pharm Sci Drug Res 2(1):48–50

Lam A, Newhouse MT (1990) Management of asthma and chronic airflow limitation: are methylxanthines obsolete? Chest 98(1):44–52

Lam S, Chan H, LeRiche JC, Chan-Yeung M, Salari H (1988) Release of leukotrienes in patients with bronchial asthma. J Allergy Clin Immunol 81(4):711–717

Lange P, Parner J, Vestbo J, Schnohr P, Jensen G (1998) A 15-year follow-up study of ventilatory function in adults with asthma. N Engl J Med 339:1194–1200

Laporte JD, Moore PE, Panettieri RA, Moeller W, Heyder J, Shore SA (1998) Prostanoids mediate IL-1beta-induced beta-adrenergic hyporesponsiveness in human airway smooth muscle cells. Am J Phys 275:L491–L501

Lee KH (2000) Research and future trends in the pharmaceutical development of medicinal herbs from Chinese medicine. Public Health Nutr 3(4a):515–522

Lee KH, Morris-Natschke S, Qian K, Dong Y, Yang X, Zhou T, Belding E, Wu SF, Wada K, Akiyama T (2012) Recent progress of research on herbal products used in traditional Chinese medicine: the herbs belonging to the divine Husbandman's herbal foundation canon (神農本草經 Shén Nóng Běn Cǎo Jīng). J Tradit Complement Med 2(1):6–26

Lee MY, Shin IS, Jeon WY, Lim HS, Kim JH, Ha H (2013) Pinellia ternata Breitenbach attenuates ovalbumin-induced allergic airway inflammation and mucus secretion in a murine model of asthma. Immunopharmacol Immunotoxicol 35(3):410–418

Levitt RC, McLane MP, MacDonald D, Ferrante V, Weiss C, Zhou T et al (1999) IL-9 pathway in asthma: new therapeutic targets for allergic inflammatory disorders. J Allergy Clin Immunol 103(5):S485–S491

Li MO, Wan YY, Flavell RA (2007) T cell-produced transforming growth factor-beta1 controls T cell tolerance and regulates Th1- and Th17-cell differentiation. Immunity 26:579–591

Li XM, Huang CK, Zhang TF, Teper AA, Srivastava K, Schofield BH, Sampson HA (2000) The Chinese herbal medicine formula MSSM-002 suppresses allergic airway hyperreactivity and modulates TH1/TH2 responses in a murine model of allergic asthma. J Allergy Clin Immunol 106(4):660–668

Lloyd CM (2007) Building better mouse models of asthma. Curr Allergy Asthma Rep 7:231–236

Lotvall J, Akdis CA, Bacharier LB, Bjermer L, Casale TB, Custovic A, Lemanske RF Jr, Wardlaw AJ, Wenzel SE, Greenberger PA (2011) Asthma endotypes: a new approach to classification of disease entities within the asthma syndrome. J Allergy Clin Immunol 127:355–360

Louahed J, Toda M, Jen J, Hamid Q, Renauld JC, Levitt RC, Nicolaides NC (2000) Interleukin-9 upregulates mucus expression in the airways. Am J Respir Cell Mol Biol 22:649–656

Louahed J, Zhou Y, Maloy WL, Rani PU, Weiss C, Tomer Y, Vink A, Renauld J, Van Snick J, Nicolaides NC, Levitt RC, Haczku A (2001) Interleukin 9 promotes influx and local maturation of eosinophils. Blood 97:1035–1042

Mak JC, Hisada T, Salmon M, Barnes PJ, Chung KF (2002) Glucocorticoids reverse IL-1beta-induced impairment of beta-adrenoceptor-mediated relaxation and up-regulation of G-protein-coupled receptor kinases. Br J Pharmacol 135:987–996

Makino T, Furuta Y, Fujii H, Nakagawa T, Wakushima H, Saito KI, Kano Y (2001) Effect of oral treatment of Perilla frutescens and its constituents on type-I allergy in mice. Biol Pharm Bull 24(10):1206–1209

Mali R, Mahajan S, Mehta A (2008) Studies on bronchodilatory effect of Lepidium sativum against allergen induced bronchospasm in Guinea pigs. Pharmacogn Mag 4(15):189

Masoli M, Fabian D, Holt S, Beasley R (2004) The global burden of asthma: executive summary of the GINA dissemination committee report. Allergy 59:469–478

Masoli M, Weatherall M, Holt S, Beasley R (2005) Moderate dose inhaled corticosteroids plus salmeterol versus higher doses of inhaled corticosteroids in symptomatic asthma. Thorax 60(9):730–734

Matera MG, Rogliani P, Cazzola M (2014) Muscarinic receptor antagonists for the treatment of chronic obstructive pulmonary disease. Expert Opin Pharmacother 15(7):961–977

Mathew A, Maclean JA, Dehaan E, Tager AM, Green FH, Luster AD (2001) Signal transducer and activator of transcription 6 controls chemokine production and T helper cell type 2 cell trafficking in allergic pulmonary inflammation. J Exp Med 193:1087–1096

Matsuda H, Tomohiro N, Ido Y, Kubo M (2002) Anti-allergic effects of cnidii monnieri fructus (dried fruits of Cnidium monnieri) and its major component, osthol. Biol Pharm Bull 25(6):809–812

McCracken JL, Tripple JW, Calhoun WJ (2016) Biologic therapy in the management of asthma. Curr Opin Allergy Clin Immunol 16(4):375–382

Mcgrath KW, Icitovic N, Boushey HA, Lazarus SC, Sutherland ER, Chinchilli VM, Fahy JV (2012) A large subgroup of mild-to-moderate asthma is persistently noneosinophilic. Am J Respir Crit Care Med 185:612–619

Mckenzie GJ, Emson CL, Bell SE, Anderson S, Fallon P, Zurawski G, Murray R, Grencis R, Mckenzie AN (1998) Impaired development of Th2 cells in IL-13-deficient mice. Immunity 9:423–432

Meena AK, Yadav A, Rao MM (2011) Ayurvedic uses and pharmacological activities of Calotropis procera Linn. Asian J Trad Med 6(2):45–53

Mei F, Xing XF, Tang QF, Chen FL, Guo Y, Song S, Tan XM, Luo JB (2016) Antipyretic and anti-asthmatic activities of traditional Chinese herb-pairs, Ephedra and gypsum. Chin J Integr Med 22(6):445–450

Mohapatra HP, Rath SP (2005) In vitro studies of Bacopa monnieri—an important medicinal plant with reference to its biochemical variations. Indian J Exp Biol 43(4):373–376

Molet S, Hamid Q, Davoine F, Nutku E, Taha R, Page N, Olivenstein R, Elias J, Chakir J (2001a) IL-17 is increased in asthmatic airways and induces human bronchial fibroblasts to produce cytokines. J Allergy Clin Immunol 108:430–438

Molet S, Hamid Q, Davoineb F, Nutku E, Tahaa R, Pagé N et al (2001b) IL-17 is increased in asthmatic airways and induces human bronchial fibroblasts to produce cytokines. J Allergy Clin Immunol 108(3):430–438

Montuschi P, Barnes PJ (2002) Exhaled leukotrienes and prostaglandins in asthma. J Allergy Clin Immunol 109(4):615–620

Morshed M, Yousefi S, Stöckle C, Simon HU, Simon D (2012) Thymic stromal lymphopoietin stimulates the formation of eosinophil extracellular traps. Allergy 67(9):1127–1137

Mosmann TR, Cherwinski H, Bond MW, Giedlin MA, Coffman RL (1986) Two types of murine helper T cell clone. I Definition according to profiles of lymphokine activities and secreted proteins. J Immunol 136:2348–2357

Mozzini Monteiro T, Ferrera Costa H, Carvalho Vieira G, Rodrigues Salgado PR, da Silva Stiebbe Salvadori MG, de Almeida RN, de Fatima Vanderlei de Souza M, Neves Matias W, Andrade Braga V, Nalivaiko E, Piuvezam MR (2016) Anti-asthmatic and anxiolytic effects of Herissantia tiubae, a Brazilian medicinal plant. Immun Inflammat Dis 4(2):201–212

Nagashima S, Yokota M, Nakai E, Kuromitsu S, Ohga K, Takeuchi M, Tsukamoto S, Ohta M (2007) Synthesis and evaluation of 2-{[2-(4-hydroxyphenyl)-ethyl]amino}pyrimidine-5-carboxamide derivatives as novel STAT6 inhibitors. Bioorg Med Chem 15:1044–1055

Nagore DH, Ghosh VK, Patil MJ (2009) Evaluation of antiasthmatic activity of Cassia sophera Linn. Pharmacogn Mag 5(19):109

Nagori BP, Solanki R (2011) Cynodon dactylon (L.) Pers.: a valuable medicinal plant. Res J Med Plant 5:508–514

Nakanishi A, Morita S, Iwashita H, Sagiya Y, Ashida Y, Shirafuji H, Fujisawa Y, Nishimura O, Fujino M (2001) Role of gob-5 in mucus overproduction and airway hyperresponsiveness in asthma. Proc Natl Acad Sci U S A 98:5175–5180

Nasser SM, Pfister R, Christie PE, Sousa AR, Barker J, Schmitz-Schumann M, Lee TH (1996) Inflammatory cell populations in bronchial biopsies from aspirin-sensitive asthmatic subjects. Am J Respir Crit Care Med 153:90–96

National Asthma Education, Prevention Program (National Heart, Lung, Blood Institute); Second Expert Panel on the Management of Asthma, National Heart, Lung and Blood Institute; National Asthma Education Program; Expert Panel on the Management of Asthma (1998) Expert panel report 2: guidelines for the diagnosis and management of asthma. National Institutes of Health, National Heart, Lung, and Blood Institute

National AE, Prevention P (2007) Expert panel report 3 (EPR-3): guidelines for the diagnosis and Management of Asthma-Summary Report 2007. J Allergy Clin Immunol 120(5 Suppl):S94

Nelson HS, Weiss ST, Bleecker ER, Yancey SW, Dorinsky PM, SMART Study Group (2006) The salmeterol multicenter asthma research trial: a comparison of usual pharmacotherapy for asthma or usual pharmacotherapy plus salmeterol. Chest 129:15–26

Nembrini C, Marsland BJ, Kopf M (2009) IL-17–producing T cells in lung immunity and inflammation. J Allergy Clin Immunol 123(5):986–994

Nguyen C, Teo JL, Matsuda A, Eguchi M, Chi EY, Henderson WR Jr, Kahn M (2003) Chemogenomic identification of Ref-1/AP-1 as a therapeutic target for asthma. Proc Natl Acad Sci U S A 100:1169–1173

Nials AT, Uddin S (2008) Mouse models of allergic asthma: acute and chronic allergen challenge. Dis Model Mech 1:213–220

Niazi J, Gupta V, Chakarborty P, Kumar P (2010) Antiinflammatory and antipyretic activity of Aleuritis moluccana leaves. Asian J Pharm Clin Res 3(1):35–37

Normansell R, Walker S, Milan SJ, Walters EH, Nair P (2014) Omalizumab for asthma in adults and children. Cochrane Database Syst Rev 1

O'byrne PM (2013) Role of monoclonal antibodies in the treatment of asthma. Can Respir J 20:23–25

O'Byrne PM, Pedersen S, Lamm CJ, Tan WC, Busse WW (2009) Severe exacerbations and decline in lung function in asthma. Am J Respir Crit Care Med 179(1):19–24

Oh CK, Geba GP, Molfino N (2010) Investigational therapeutics targeting the IL-4/IL-13/STAT-6 pathway for the treatment of asthma. Eur Respir Rev 19:46–54

Oh CK, Leigh R, Mclaurin KK, Kim K, Hultquist M, Molfino NA (2013) A randomized, controlled trial to evaluate the effect of an anti-interleukin-9 monoclonal antibody in adults with uncontrolled asthma. Respir Res 14:93

Oh CK, Raible D, Geba GP, Molfino NA (2011) Biology of the interleukin-9 pathway and its therapeutic potential for the treatment of asthma. Inflamm Allergy Drug Targets 10:180–186

Okpo SO, Fatokun F, Adeyemi OO (2001) Analgesic and anti-inflammatory activity of Crinum glaucum aqueous extract. J Ethnopharmacol 78(2–3):207–211

Ordonez CL, Khashayar R, Wong HH, Ferrando R, Wu R, Hyde DM, Hotchkiss JA, Zhang Y, Novikov A, Dolganov G, Fahy JV (2001) Mild and moderate asthma is associated with airway goblet cell hyperplasia and abnormalities in mucin gene expression. Am J Respir Crit Care Med 163:517–523

Organization WH (1990) The use of traditional medicine in primary health care. A manual for health Workers in South-East Asia. SEARO Regional Health Papers; New Delhi

Ortega HG, Liu MC, Pavord ID, Brusselle GG, FitzGerald JM, Chetta A et al (2014) Mepolizumab treatment in patients with severe eosinophilic asthma. N Engl J Med 371(13):1198–1207

Ouyang W, Lohning M, Gao Z, Assenmacher M, Ranganath S, Radbruch A, Murphy KM (2000) Stat6-independent GATA-3 autoactivation directs IL-4-independent Th2 development and commitment. Immunity 12:27–37

Ouyang W, Ranganath SH, Weindel K, Bhattacharya D, Murphy TL, SHA WC, Murphy KM (1998) Inhibition of Th1 development mediated by GATA-3 through an IL-4-independent mechanism. Immunity 9:745–755

Padilla J, Daley E, Chow A, Robinson K, Parthasarathi K, McKenzie AN et al (2005) IL-13 regulates the immune response to inhaled antigens. J Immunol 174(12):8097–8105

Pai SY, Truitt ML, Ho IC (2004) GATA-3 deficiency abrogates the development and maintenance of T helper type 2 cells. Proc Natl Acad Sci U S A 101:1993–1998

Paliwa JK, Dwivedi AK, Singh S, Gutpa RC (2000) Pharmacokinetics and in-situ absorption studies of a new anti-allergic compound 73/602 in rats. Int J Pharm 197(1–2):213–220

Pandey MM, Rastogi S, Rawat AKS (2007) Saussurea costus: botanical, chemical and pharmacological review of an ayurvedic medicinal plant. J Ethnopharmacol 110(3):379–390

Parker JM, Glaspole IN, Lancaster LH, Haddad TJ, She D, Roseti SL et al (2018) A phase 2 randomized controlled study of tralokinumab in subjects with idiopathic pulmonary fibrosis. Am J Respir Crit Care Med 197(1):94–103

Parker JM, Oh CK, Laforce C, Miller SD, Pearlman DS, Le C, Robbie GJ, White WI, White B, Molfino NA (2011a) Safety profile and clinical activity of multiple subcutaneous doses of MEDI-528, a humanized anti-interleukin-9 monoclonal antibody, in two randomized phase 2a studies in subjects with asthma. BMC Pulm Med 11:14

Parker JM, Oh CK, LaForce C, Miller SD, Pearlman DS, Le C et al (2011b) Safety profile and clinical activity of multiple subcutaneous doses of MEDI-528, a humanized anti-interleukin-9 monoclonal antibody, in two randomized phase 2a studies in subjects with asthma. BMC Pulm Med 11(1):14

Patel KG, Rao NJ, Gajera VG, Bhatt PA, Patel KV, Gandhi TR (2010) Anti-allergic activity of stem bark of Myrica esculenta Buch.-ham.(Myricaceae). J Young Pharm 2(1):74–78

Patel S, Patel V (2019) Inhibitory effects of catechin isolated from Acacia catechu on ovalbumin induced allergic asthma model. Nutr Food Sci 49(1):18–31

Peng Q, Matsuda T, Hirst SJ (2004) Signaling pathways regulating interleukin-13-stimulated chemokine release from airway smooth muscle. Am J Respir Crit Care Med 169:596–603

Petrovic-Stanojevic N, Topic A, Nikolic A, Stankovic M, Dopudja-Pantic V, Milenkovic B, Radojkovic D (2014) Polymorphisms of Beta2-adrenergic receptor gene in Serbian asthmatic adults: effects on response to Beta-agonists. Mol Diagn Ther 18(6):639–646

Piper E, Brightling C, Niven R, Oh C, Faggioni R, Poon K et al (2013) A phase II placebo-controlled study of tralokinumab in moderate-to-severe asthma. Eur Respir J 41(2):330–338

Prasad R, Lawania RD, Gupta R (2009) Role of herbs in the management of asthma. Pharmacogn Rev 3(6):247

Quirce S, Bobolea I, Dominguez-ortega J, Barranco P (2014) Future biologic therapies in asthma. Arch Bronconeumol 50:355–361

Ram A, Mabalirajan U, Das M, Bhattacharya I, Dinda AK, Gangal SV, Ghosh B (2006) Glycyrrhizin alleviates experimental allergic asthma in mice. Int Immunopharmacol 6(9):1468–1477

Ram FSF, Cates CJ, Ducharme FM (2005) Long-acting beta2-agonists versus anti-leukotrienes as add-on therapy to inhaled corticosteroids for chronic asthma. Cochrane Database Syst Rev;(1):CD003137

Ramalingam TR, Pesce JT, Sheikh F, Cheever AW, Mentink-Kane MM, Wilson MS, Stevens S, Valenzuela DM, Murphy AJ, Yancopoulos GD, Urban JF Jr, Donnelly RP, Wynn TA (2008) Unique functions of the type II interleukin 4 receptor identified in mice lacking the interleukin 13 receptor alpha1 chain. Nat Immunol 9:25–33

Ramanitrahasimbola D, Rakotondramanana DA, Rasoanaivo P, Randriantsoa A, Ratsimamanga S, Palazzino G, Galeffi C, Nicoletti M (2005) Bronchodilator activity of Phymatodes scolopendria (Burm.) Ching and its bioactive constituent. J Ethnopharmacol 102(3):400–407

Rani AS, Patnaik S, Sulakshanaand G, Saidulu B (2012) Review of Tylophora indica-an antiasthmatic plant. FS J Res Basic Appl Sci 1(3):20–21

Ranjeeta P, Lawania RD, Rajiv G (2009) Role of herbs in the management of asthma. Pharmacogn Rev 3(6):247–258

Ray A, Cohn L (1999) Th2 cells and GATA-3 in asthma: new insights into the regulation of airway inflammation. J Clin Invest 104:985–993

Rayees S, Malik F, Bukhari SI, Singh G (2014) Linking GATA-3 and interleukin-13: implications in asthma. Inflamm Res 63:255–265

Redington AE (2000) Fibrosis and airway remodelling. Clin Exp Allergy 30(Suppl 1):42–45

Robinson DS, Campbell D, Barnes PJ (2001) Addition of leukotriene antagonists to therapy in chronic persistent asthma: a randomised double-blind placebo-controlled trial. Lancet 357(9273):2007–2011

Robinson DS, Hamid Q, Ying S, Tsicopoulos A, Barkans J, Bentley AM, Corrigan C, Durham SR, Kay AB (1992) Predominant TH2-like bronchoalveolar T-lymphocyte population in atopic asthma. N Engl J Med 326:298–304

Rodrigo GJ, Rodrigo C (2002) The role of anticholinergics in acute asthma treatment: an evidence-based evaluation. Chest 121(6):1977–1987

Roland NJ, Bhalla RK, Earis J (2004) The local side effects of inhaled corticosteroids: current understanding and review of the literature. Chest 126(1):213–219

Rowe BH, Spooner C, Ducharme F, Bretzlaff J, Bota G (2001) Early emergency department treatment of acute asthma with systemic corticosteroids. Cochrane Database Syst Rev 1

Russell RJ, Chachi L, FitzGerald JM, Backer V, Olivenstein R, Titlestad IL, Ulrik CS, Harrison T, Singh D, Chaudhuri R, Leaker B, McGarvey L, Siddiqui S, Wang M, Braddock M, Nordenmark LH, Cohen D, Parikh H, Colice G, Brightling CE, MESOS study investigators. (2018) Effect of tralokinumab, an interleukin-13 neutralising monoclonal antibody, on eosinophilic airway inflammation in uncontrolled moderate-to-severe asthma (MESOS): a multicentre, double-blind, randomised, placebo-controlled phase 2 trial. Lancet Respir Med 6(7):499–510

Ryu MH, Jha A, Ojo OO, Mahood TH, Basu S, Detillieux KA, Nikoobakht N, Wong CS, Loewen M, Becker AB, Halayko AJ (2014) Chronic exposure to perfluorinated compounds: impact on airway hyperresponsiveness and inflammation. Am J Physiol Lung Cell Mol Physiol

Saganuwan AS (2010) Some medicinal plants of Arabian Pennisula. J Med Plants Res 4(9):766–788

Sampson AP, Cowburn AS, Sladek K, Adamek L, Nizankowska E, Szczeklik A, Lam BK, Penrose JF, Austen KF, Holgate ST (1997) Profound overexpression of leukotriene C4 synthase in bronchial biopsies from aspirin-intolerant asthmatic patients. Int Arch Allergy Immunol 113:355–357

Santini G, Mores N, Malerba M, Mondino C, Anzivino R, Macis G, Montuschi P (2017) Dupilumab for the treatment of asthma. Expert Opin Investig Drugs 26(3):357–366

Sarkar BK, Solanki SS (2011) Isolation, characterization and antibacterial activity of leaves extract of bael (Aegle marmelos). Int J Pharm Life Sci 2(12):1303–1305

Sarpate RV, Deore TK, Tupkari SV (2009) Bronchodilatory effect of Sphaeranthus indicus Linn against allergen induced bronchospasm in Guinea pigs. Pharmacogn Mag 5(19):74

Sarpong SB, Zhang LY, Kleeberger SR (2003) A novel mouse model of experimental asthma. Int Arch Allergy Immunol 132:346–354

Savithramma N, Sulochana C, Rao KN (2007) Ethnobotanical survey of plants used to treat asthma in Andhra Pradesh, India. J Ethnopharmacol 113(1):54–61

Saxena P, Saxena P (2014) In-vitro and in-vivo evaluation of anti asthmatic activity of rhizomes extract of Acorus calamus (Linn.) in Guinea pigs. Res J Pharm Sci 3(5):1–6

Sehgal R, Chauhan A, Gilhotra UK, Gilhotra A (2013) In vitro and in-vivo evaluation of antiasthmatic activity of *Picrorhiza kurroa* plant. Int J Pharm Sci Res 4(9):3440

Selroos O, Pietinalho A, Löfroos AB, Riska H (1995) Effect of early vs late intervention with inhaled corticosteroids in asthma. Chest 108(5):1228–1234

Sestini P, Renzoni E, Robinson S, Poole P, Ram FS (2002) Short-acting beta2-agonists for stable chronic obstructive pulmonary disease. Cochrane Database Syst Rev (3)

Shakarami Z, Ghaleh HEG, Motlagh BM, Sheikhian A, Kondori BJ (2019) Evaluation of the protective and therapeutic effects of Pistacia atlantica gum aqueous extract on cellular and pathological aspects of experimental asthma in Balb/c mice. Avicenna J Phytomed 9(3):248

Shang JH, Cai XH, Zhao YL, Feng T, Luo XD (2010) Pharmacological evaluation of Alstonia scholaris: anti-tussive, anti-asthmatic and expectorant activities. J Ethnopharmacol 129(3):293–298

Shao CR, Chen FM, Tang YX (1994) Clinical and experimental study on Ligusticum wallichii mixture in preventing and treating bronchial asthma. Zhongguo Zhong xi yi jie he za zhi Zhongguo Zhongxiyi jiehe zazhi= Chin J Integr Trad West Med 14(8):465–468

Shikotra A, Choy DF, Ohri CM, Doran E, Butler C, Hargadon B, Shelley M, Abbas AR, Austin CD, Jackman J, Wu LC, Heaney LG, Arron JR, Bradding P (2012) Increased expression of immunoreactive thymic stromal lymphopoietin in patients with severe asthma. J Allergy Clin Immunol 129(104–11):e1–e9

Shimoda K, Van Deursen J, Sangster MY, Sarawar SR, Carson RT, Tripp RA, Chu C, Quelle FW, Nosaka T, Vignali DA, Doherty PC, Grosveld G, Paul WE, Ihle JN (1996) Lack of IL-4-induced Th2 response and IgE class switching in mice with disrupted Stat6 gene. Nature 380:630–633

Shin YS, Takeda K, Gelfand EW (2009) Understanding asthma using animal models. Allergy Asthma Immunol Res 1:10–18

Shindo K, Fukumura M, Miyakawa K (1994) Plasma levels of leukotriene E4 during clinical course of bronchial asthma and the effect of oral prednisolone. Chest 105:1038–1041

Silva RA, Almeida FM, Olivo CR, Saraiva-Romanholo BM, Martins MA, Carvalho CR (2014) Airway remodeling is reversed by aerobic training in a murine model of chronic asthma. Scand J Med Sci Sports

Sim Y, Park G, Eo H, Huh E, Gu PS, Hong SP, Pak YK, Oh MS (2017) Protective effects of a herbal extract combination of Bupleurum falcatum, Paeonia suffruticosa, and Angelica dahurica against MPTP-induced neurotoxicity via regulation of nuclear receptor-related 1 protein. Neuroscience 340:166–175

Singh A, Handa SS (1995) Hepatoprotective activity of Apium graveolens and Hygrophila auriculata against paracetamol and thioacetamide intoxication in rats. J Ethnopharmacol 49(3):119–126

Singh G, Sharma PK, Dudhe R, Singh S (2010) Biological activities of Withania somnifera. Ann Biol Res 1(3):56–63

Singh RK, Acharya SB, Bhattacharya SK (2000) Pharmacological activity of Elaeocarpus sphaericus. Phytother Res 14(1):36–39

Singh RP, Gangadharappa HV, Mruthunjaya K (2017) Cuminum cyminum–a popular spice: an updated review. Pharm J 9(3)

Singh SK, Patel JR, Dubey PK, Thakur S (2014) A review on antiasthmatic activity of traditional medicinal plants. Int J Pharm Sci Res 5(10):4109–4116

Smiroldo TL (2005) A complementary approach to asthma in paediatrics. Aust J Med Herbalism 17(1):15

Sokol CL, Barton GM, Farr AG, Medzhitov R (2008) A mechanism for the initiation of allergen-induced T helper type 2 responses. Nat Immunol 9:310–318

Soler M, Matz J, Townley R, Buhl R, O'brien J, Fox H et al (2001) The anti-IgE antibody omalizumab reduces exacerbations and steroid requirement in allergic asthmatics. Eur Respir J 18(2):254–261

Sporik R, Ingram JM, Price W, Sussman JH, Honsinger RW, Platts-Mills TA (1995) Association of asthma with serum IgE and skin test reactivity to allergens among children living at high altitude. Tickling the dragon's breath. Am J Respir Crit Care Med 151:1388–1392

Srivastava KD, Zou ZM, Sampson HA, Dansky H, Li XM (2005) Direct modulation of airway reactivity by the Chinese anti-asthma herbal formula ASHMI. J Allergy Clin Immunol 115(2):S7

Stansbury J, Saunders PR, Zampieron E (2013) The use of lobelia in the treatment of asthma and respiratory illness. J Restorat Med 2(1):94–100

Stein ML, Collins MH, Villanueva JM, Kushner JP, Putnam PE, Buckmeier BK et al (2006) Anti–IL-5 (mepolizumab) therapy for eosinophilic esophagitis. J Allergy Clin Immunol 118(6):1312–1319

Suissa S, Ernst P (2001) Inhaled corticosteroids: impact on asthma morbidity and mortality. J Allergy Clin Immunol 107(6):937–944

Suissa S, Ernst P, Benayoun S, Baltzan M, Cai B (2000) Low-dose inhaled corticosteroids and the prevention of death from asthma. N Engl J Med 343(5):332–336

Taube C, Dakhama A, Gelfand EW (2004) Insights into the pathogenesis of asthma utilizing murine models. Int Arch Allergy Immunol 135:173–186

Taur DJ, Patil RY (2011) Some medicinal plants with antiasthmatic potential: a current status. Asian Pac J Trop Biomed 1(5):413–418

Tayade PM, Ghaisas MM, Jagtap SA, Dongre SH (2009) Anti-asthmatic activity of methanolic extract of leaves of Tamarindus Indica Linn. J Pharm Res 2(5):944–947

Toward TJ, Broadley KJ (2002) Goblet cell hyperplasia, airway function, and leukocyte infiltration after chronic lipopolysaccharide exposure in conscious Guinea pigs: effects of rolipram and dexamethasone. J Pharmacol Exp Ther 302:814–821

Trinchieri G, Sher A (2007) Cooperation of toll-like receptor signals in innate immune defence. Nat Rev Immunol 7:179–190

Trinchieri G, Pflanz S, Kastelein RA (2003) The IL-12 family of heterodimeric cytokines: new players in the regulation of T cell responses. Immunity 19:641–644

Ulrik CS, Lange P (1994) Decline of lung function in adults with bronchial asthma. Am J Respir Crit Care Med 150:629–634

Urata Y, Yoshida S, Irie Y, Tanigawa T, Amayasu H, Nakabayashi M, Akahori K (2002) Treatment of asthma patients with herbal medicine TJ-96: a randomized controlled trial. Respir Med 96(6):469–474

Usui T, Nishikomori R, Kitani A, Strober W (2003) GATA-3 suppresses Th1 development by downregulation of Stat4 and not through effects on IL-12Rbeta2 chain or T-bet. Immunity 18:415–428

Vashisth A, Singh R, Kakar S (2017) Asthma and medicinal plants: A review. Int J Recent Adv Sci Technol 4(4):1–7

Vatrella A, Fabozzi I, Calabrese C, Maselli R, Pelaia G (2014) Dupilumab: a novel treatment for asthma. J Asthma Allergy 7:123

Veldhoen M, Hocking RJ, Atkins CJ, Locksley RM, Stockinger B (2006) TGFβ in the context of an inflammatory cytokine milieu supports de novo differentiation of IL-17-producing T cells. Immunity 24:179–189

Verma SL, Shrivastava A (2015) Popular herbs of Chhatisgarh and their uses in the treatment of common diseases in baster region. Indian J Life Sci 5(1):129–133

Vernon MK, Wiklund I, Bell JA, Dale P, Chapman KR (2012) What do we know about asthma triggers? A review of the literature. J Asthma 49(10):991–998

Vezina K, Chauhan BF, Ducharme FM (2014) Inhaled anticholinergics and short-acting beta 2-agonists versus short-acting beta2-agonists alone for children with acute asthma in hospital. Cochrane Database of Systematic Reviews;(7):CD010283

Von Bulow A, Backer V, Porsbjerg C (2014) Severe asthma - where are we today? Ugeskr Laeger 176(3A):V05130307

Walsh GM (2009) Reslizumab, a humanized anti-IL-5 mAb for the treatment of eosinophil-mediated inflammatory conditions. Curr Opin Mol Ther 11:329–336

Walsh GM (2011) Novel cytokine-directed therapies for asthma. Discov Med 11:283–291

Walsh GM (2013a) An update on biologic-based therapy in asthma. Immunotherapy 5:1255–1264

Walsh GM (2013b) Profile of reslizumab in eosinophilic disease and its potential in the treatment of poorly controlled eosinophilic asthma. Biologics Targets Ther 7:7

Walsh GM, Al-Rabia M, Blaylock MG, Sexton DW, Duncan CJ, Lawrie A (2005) Control of eosinophil toxicity in the lung. Curr Drug Targets Inflamm Allergy 4:481–486

Walters EH, Walters JA, Gibson PG, Jones P (2003) Inhaled short acting beta2-agonist use in chronic asthma: regular versus as needed treatment. Cochrane Database Syst Rev;(1)

Wanner A, Salathe M, O'riordan TG (1996) Mucociliary clearance in the airways. Am J Respir Crit Care Med 154:1868–1902

Webb DC, Mckenzie AN, Koskinen AM, Yang M, Mattes J, Foster PS (2000) Integrated signals between IL-13, IL-4, and IL-5 regulate airways hyperreactivity. J Immunol 165:108–113

Wechsler ME (2013) Inhibiting interleukin-4 and interleukin-13 in difficult-to-control asthma. N Engl J Med 368(26):2511–2513

Wen MC, Wei CH, Hu ZQ, Srivastava K, Ko J, Xi ST, Mu DZ, Du JB, Li GH, Wallenstein S, Sampson H (2005) Efficacy and tolerability of antiasthma herbal medicine intervention in adult patients with moderate-severe allergic asthma. J Allergy Clin Immunol 116(3):517–524

Westby MJ, Benson MK, Gibson PG (2004) Anticholinergic agents for chronic asthma in adults. Cochrane Database Syst Rev 3

Wills-Karp M, Karp CL (2004) Biomedicine. Eosinophils in asthma: remodeling a tangled tale. Science 305:1726–1729

Wills-Karp M, Luyimbazi J, Xu X, Schofield B, Neben TY, Karp CL, Donaldson DD (1998) Interleukin-13: central mediator of allergic asthma. Science 282:2258–2261

Woolley MJ, Denburg JA, Ellis R, Dahlback M, O'Byrne PM (1994) Allergen-induced changes in bone marrow progenitors and airway responsiveness in dogs and the effect of inhaled budesonide on these parameters. Am J Respir Cell Mol Biol 11(5):600–606

Wu QZ, Zhao DX, Xiang J, Zhang M, Zhang CF, Xu XH (2016) Antitussive, expectorant, and anti-inflammatory activities of four caffeoylquinic acids isolated from Tussilago farfara. Pharm Biol 54(7):1117–1124

Yadav SA, Ramalingam S, Raj AJ, Subban R (2015) Antihistamine from Tragia involucrata L. leaves. J Complement Integrat Med 12(3):217–226

Youssouf MS, Kaiser P, Tahir M, Singh GD, Singh S, Sharma VK, Satti NK, Haque SE, Johri RK (2007) Anti-anaphylactic effect of Euphorbia hirta. Fitoterapia 78(7–8):535–539

Zahidin NS, Saidin S, Zulkifli RM, Muhamad II, Ya'akob H, Nur H (2017) A review of Acalypha indica L.(Euphorbiaceae) as traditional medicinal plant and its therapeutic potential. J Ethnopharmacol 207:146–173

Zhang T, Srivastava K, Wen MC, Yang N, Cao J, Busse P, Birmingham N, Goldfarb J, Li XM (2010) Pharmacology and immunological actions of a herbal medicine ASHMI™ on allergic asthma. Phytother Res 24(7):1047–1055

Zhou E, Fu Y, Wei Z, Yang Z (2014) Inhibition of allergic airway inflammation through the blockage of NF-kappaB activation by ellagic acid in an ovalbumin-induced mouse asthma model. Food Funct 5:2106–2112

Zhou M, Ouyang W (2003) The function role of GATA-3 in Th1 and Th2 differentiation. Immunol Res 28:25–37

Zhu J, Guo L, Watson CJ, Hu-Li J, Paul WE (2001) Stat6 is necessary and sufficient for IL-4's role in Th2 differentiation and cell expansion. J Immunol 166:7276–7281

Zhu J, Min B, Hu-Li J, Watson CJ, Grinberg A, Wang Q, Killeen N, Urban JF Jr, Guo L, Paul WE (2004) Conditional deletion of Gata3 shows its essential function in T(H)1-T(H)2 responses. Nat Immunol 5:1157–1165

Zhu J, Yamane H, Cote-Sierra J, Guo L, Paul WE (2006) GATA-3 promotes Th2 responses through three different mechanisms: induction of Th2 cytokine production, selective growth of Th2 cells and inhibition of Th1 cell-specific factors. Cell Res 16:3–10

Zhu Z, Homer RJ, Wang Z, Chen Q, Geba GP, Wang J, Zhang Y, Elias JA (1999) Pulmonary expression of interleukin-13 causes inflammation, mucus hypersecretion, subepithelial fibrosis, physiologic abnormalities, and eotaxin production. J Clin Invest 103:779–788

Ziegler SF, Roan F, Bell BD, Stoklasek TA, Kitajima M, Han H (2013) The biology of thymic stromal lymphopoietin (TSLP). Adv Pharmacol 66:129–155. Academic

Zosky GR, Sly PD (2007) Animal models of asthma. Clin Exp Allergy 37:973–988

Printed in the United States
by Baker & Taylor Publisher Services